普通高等院校土木专业“十四五”规划精品教材

土木工程专业英语

English for Civil Engineering

（第三版）

丛书审定委员会

王思敬　彭少民　石永久　白国良

李　杰　姜忻良　吴瑞麟　张智慧

本书主审　单　建

本书主编　秦卫红

本书副主编

陈昌平　许婷华　董　平

本书编写委员会

秦卫红　陈昌平　许婷华　董　平

张　晋　戴国亮　陈静茹　张明慧

华中科技大学出版社

中国·武汉

内 容 提 要

全书内容分为建筑与结构力学、建筑与管理工程、桥梁和道路工程、地基基础、港口工程等几个部分。每部分选材涵盖该专业所涉及的主要内容，绝大多数原文作者母语为英语，每篇力求做到语言表达地道，不出现生涩词汇。

教学模块及特点：课文正文后面附有专业词汇（带有音标）和常用词组，以及部分句子或内容的注释和说明，以便于读者快速、及时、准确掌握文章内容；课后所附相关练习有词汇或词组的中英文互译、句子的中英文互译和问答题等题型，以供课堂教学或者自学时及时考查对课文的掌握程度之用。每篇正文后面均有一篇与正文内容相关的阅读材料。正文和阅读材料不设全篇译文，避免产生中文依赖。全书以较小的篇幅介绍了土木工程专业英语中常用的阅读和写作技巧，并将内容由浅到深地分配到每一课课后，希望能借此提高读者的英语阅读和写作水平。

本书适合普通高等院校土木工程专业本科生使用，也可供土木工程师及相关从业人员阅读参考，对高校教师和研究生也有参考价值。

图书在版编目(CIP)数据

土木工程专业英语/秦卫红主编. —3 版. —武汉：华中科技大学出版社，2020.8 (2024.8 重印)
ISBN 978-7-5680-6338-8

Ⅰ. ①土… Ⅱ. ①秦… Ⅲ. ①土木工程-英语-教材 Ⅳ. ①TU

中国版本图书馆 CIP 数据核字(2020)第 134987 号

土木工程专业英语(第三版) 秦卫红 主编
Tumu Gongcheng Zhuanye Yingyu (Di-san Ban)

策划编辑：简晓思 责任编辑：简晓思 梁 任
封面设计：张 璐 责任校对：李 琴
责任监印：朱 玢
出版发行：华中科技大学出版社(中国·武汉) 电话：(027)81321913
武汉市东湖新技术开发区华工科技园 邮编：430223
录 排：武汉楚海文化传播有限公司
印 刷：武汉科源印刷设计有限公司
开 本：850mm×1065mm 1/16
印 张：18.25
字 数：399 千字
版 次：2024 年 8 月第 3 版第 3 次印刷
定 价：59.80 元

普通高等院校土木专业“十四五”规划精品教材

总 序

教育可理解为教书与育人。所谓教书，不外乎是教给学生科学知识、技术方法和运作技能等，教学生以安身之本。所谓育人，则要教给学生做人的道理，提升学生的人文素质和科学精神，教学生以立命之本。我们教育工作者应该从中华民族振兴的历史使命出发，来从事教书与育人工作。作为教育本源之一的教材，必然要承载教书和育人的双重责任，体现两者的高度结合。

中国经济建设高速持续发展，国家对各类建筑人才需求日增，对高校土建类高素质人才培养提出了新的要求，从而对土建类教材建设也提出了新的要求。这套教材正是为了适应当今时代对高层次建设人才培养的需求而编写的。

一部好的教材应该把人文素质和科学精神的培养放在重要位置。教材中不仅要从内容上体现人文素质教育和科学精神教育，而且要从科学严谨性、法规权威性、工程技术创新性来启发和促进学生科学世界观的形成。简而言之，这套教材有以下特点。

(1)从指导思想来讲，这套教材注意到“六个面向”，即面向社会需求、面向建筑实践、面向人才市场、面向教学改革、面向学生现状、面向新兴技术。

(2)教材编写体系有所创新。结合具有土建类学科特色的教学理论、教学方法和教学模式，这套教材进行了许多新的教学方式的探索，如引入案例式教学、研讨式教学等。

(3)这套教材适应现在教学改革发展的要求，提倡所谓“宽口径、少学时”的人才培养模式。在教学体系、教材编写内容和数量等方面也做了相应改变，而且教学起点也可随着学生水平做相应调整。同时，在这套教材编写中，特别重视人才的能力培养和基本技能培养，适应土建专业特别强调实践性的要求。

我们希望这套教材能有助于培养适应社会发展需要的、素质全面的新型工程建设人才。我们也相信这套教材能达到这个目标，从形式到内容都成为精品，为教师和学生以及专业人士所喜爱。

中国工程院院士 王思敬

2006年6月于北京

第三版前言

《土木工程专业英语》作为一本面向土木工程专业设置的专业英语类教材，自2009年9月出版，2012年再版以来，已在多所高等院校使用，并受到广大师生的好评与欢迎。

选用本书的师生在基本认可其内容和形式的同时，也给编者带来了宝贵的反馈意见，因此，我们根据学科发展，并针对培养对象，对本教材进行修订。

本书作者都具备在高等院校从事专业英语教学的经验，根据教学实践对本书再版的内容确定如下。

第三版仍采用第一版的基本框架和基本内容。本次修订除全面校正第一版中的欠妥之处外，还重点修订了“土木工程专业英语论文写作技巧”的内容。

本书适合如下专业的读者使用：建筑工程、桥梁工程、道路工程、地下工程、港口工程等。

本书在修订过程中，虽然编者力求严谨和正确，但限于学识水平与能力，书中不足之处仍属难免，殷切希望读者批评指正。同时，对在修订工作中给予大力支持的本书编委会的各位专家、出版社编辑表示深深的谢意。

本书出版后的这段时间，作者收到很多读者的电子邮件，对本书的写作团体提供了宝贵的意见，在第三版出版之际向这些热心读者表示衷心的感谢！希望每一位读者都能继续关注本书的发展和完善，继续提出您的意见和建议！

秦卫红

2020年8月

qinweihongseu@163.com

目　　录

Lesson 1

STRESS AND STRAIN

The concepts of stress and strain can be illustrated in an elementary way by considering the extension of a prismatic bar (see Fig. 1-1(a)). A prismatic bar is one that has constant cross section through its length and a straight axis. In this illustration, the bar is assumed to be loaded at its ends by axial forces P that produce a uniform stretching, or tension, of the bar. By making an artificial cut (section m-m) through the bar at right angles to its axis, we can isolate parts of the bar as a free body (see Fig. 1-1(b)). ① At the right-hand end the tensile force P is applied, and at the other end, these forces will be continuously distributed over the cross section, analogous to the continuous distribution of hydrostatic pressure over a submerged surface. ② The intensity of force, that is, the force per unit area is called the stress and is commonly denoted by the Greek letter σ. Assuming that the stress has a uniform distribution over the cross section (see Fig. 1-1(b)), we can readily see that its resultant is equal to the intensity σ times the cross-sectional area A of the bar. Furthermore, from the equilibrium of the body shown in Fig. 1-1(b), we can also see that this resultant must be equal in magnitude and opposite in direction to the force P. Hence, we obtain

$$\sigma=\frac{P}{A} \qquad 1\text{-}1$$

as the equation for the uniform stress in a prismatic bar. This equation shows that stress has units of force divided by area—for example, pounds per square inch (psi) or kips* per square inch (ksi). When the bar is being stretched by the force P, as shown in the figure, the resulting stress is a tensile stress; if the forces are reversed in direction, causing the bar to be compressed, they are called compressive

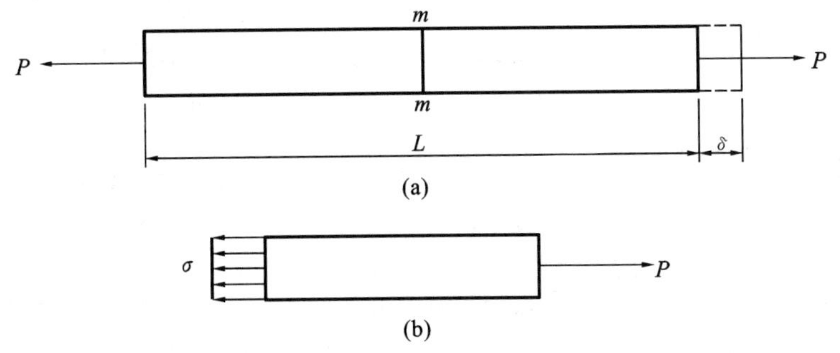

Fig. 1-1 Prismatic bar in tension

stresses.

A necessary condition for Eq. 1-1 to be valid is that the stress σ must be uniform over the cross section of the bar. ③ This condition will be realized if the axial force P acts through the centroid of the cross section, as can be demonstrated by statics. ④ When the load P does not act at the centroid, bending of the bar will result, and a more complicated analysis is necessary. However, here, it is assumed that all axial forces are applied at the centroid of the cross section unless specifically stated to the contrary. Also, unless stated otherwise, ⑤ it is generally assumed that the weight of the object itself is neglected, as was done when discussing the bar in Fig. 1-1.

The total elongation of a bar carrying an axial force will be denoted by the Greek letter δ (see Fig. 1-1(a)), and the elongation per unit length, or strain, is then determined by the equation

$$\varepsilon=\frac{\delta}{L} \tag{1-2}$$

where L is the total length of the bar. Note that the strain ε is a nondimensional quantity. It can be obtained accurately from Eq. 1-2 as long as the strain is uniform throughout the length of the bar. If the bar is in tension, the strain is a tensile strain, representing an elongation or stretching of the material; if the bar is in compression, the strain is a compressive strain, which means that adjacent across sections of the bar move closer to one another.

* One kip, or kilopound, equals 1000 pounds.

Words and Expressions

stress	[stres]	*n.* 应力
strain	[strein]	*n.* 应变
prismatic	[priz'mætik]	*adj.* 棱柱形的,棱镜的
load	[ləud]	*n.* 负荷,荷载,负载 *v.* 加载
uniform	['juːnifɔːm]	*adj.* 均匀的
tension	['tenʃən]	*n.* 张力,拉力
tensile	['tensail]	*adj.* 张力的,拉伸的
distribute	[di'stribjuːt]	*v.* 分布,区分
analogous	[ə'næləgəs]	*adj.* 类似的,相似的
distribution	[ˌdistri'bjuːʃən]	*n.* 分布
hydrostatic	[ˌhaidrəu'stætik]	*adj.* 流体静力学的
submerge	[səb'mɜːdʒ]	*v.* 浸没,淹没
intensity	[in'tensəti]	*n.* 强度

denote	[di'nəut]	*v.* 表示,代表
resultant	[ri'zʌltənt]	*n.* 结果,合力 *adj.* 合成的,组合的
equilibrium	[ˌi:kwi'libriəm]	*n.* 平衡,均衡
statics	['stætiks]	*n.* 静力学
centroid	['sentrɔid]	*n.* 重心,质心,形心(曲线)
elongation	[ˌi:lɔŋ'geiʃən]	*n.* 伸长,延长(部分)
nondimensional	['nɔndi'menʃənəl]	*adj.* 无量纲的
compression	[kəm'preʃən]	*n.* 压缩
compressive	[kəm'presiv]	*adj.* 有压力的,压缩的
adjacent	[ə'dʒeisənt]	*adj.* 附近的
in an elementary way		以基本方法
constant cross section		等截面
axial force		轴向力
at right angles		成直角
be analogous to		类似于
tensile stress		拉应力
compressive stress		压应力
to the contrary		意思相反的(地)
be in tension		受拉
tensile strain		拉应变
be in compression		受压

Notes

① 本课和书中以后各课出现的-ing 形式(包括传统语法的现在分词和动名词),在句中它们常作宾语和状语。
第一段中出现了 by considering,by making。-ing 形式作介词宾语,它与介词在一起构成介词短语,在句中作状语。第二段最后一句出现了 when discussing the bar in Fig. 1-1,由连词 when 引起的-ing 分词短语在句中作时间状语。

② these forces will be continuously distributed over the cross section, analogous to the continuous distribution of hydrostatic pressure over a submerged surface.
句中 analogous to 是形容词短语作说明语。全句意思是:这些力连续分布在整个横截面上,类似于浸没面上液体静压力的连续分布。

③ A necessary condition for Eq. 1-1 to be valid is that the stress σ must be uniform over the cross section of the bar.
for Eq. 1-1 是动词不定式 to be valid 的逻辑主语,意思为:公式 1-1 能够成立的必要条件是……。后面的 that 引出一个表语从句。

④ This condition will be realized if the axial force P acts through the centroid of the cross section, as can be demonstrated by statics.

as can be demonstrated 及本段最后一句中的 as was done,其中的 as 是关系代词,引出非限制性定语从句,as 代表整个主句所讲的内容,并在从句中作主语。

⑤ However, here, it is assumed that all axial forces are applied at the centroid of the cross section unless specifically stated to the contrary. Also, unless stated otherwise…

前一句中由 unless 引导的从句意为:除非特殊说明不同情况。后一句:同样,除非另有说明……

Exercises

Ⅰ. Translate the following words into Chinese.

1. the total elongation of a bar carrying an axial force ____________
2. a bar that has constant cross section throughout its length and a strait axis ____________
3. a uniform stretching ____________
4. units of force divided by area ____________
5. a compressive stress ____________
6. at right angles to its axis ____________

Ⅱ. Translate the following words into English.

1. 等截面____________
2. 单位面积上的力____________
3. 无量纲的量____________
4. 杆件的横截面面积____________
5. 横截面的形心____________
6. 单位长度伸长量____________
7. 相邻横截面____________
8. 除非另有说明____________

Ⅲ. Translate the following sentences into Chinese.

1. This condition will be realized if the axial force P acts through the centroid of the cross section, as can be demonstrated by statics.
2. Also, unless stated otherwise, it is generally assumed that the weight of the object itself is neglected, as was done when discussing the bar in Fig. 1-1.
3. By making an artificial cut through the bar at right angles to its axis, we can isolate parts of the bar as a free body.
4. The concepts of stress and strain can be illustrated in an elementary way by considering the extension of a prismatic bar.
5. Assuming that the stress has a uniform distribution over the cross section, we can readily see that its resultant is equal to the intensity σ times the cross-sec-

tional area A of the bar.

6. When the bar is being stretched by the force P, as shown in the figure, the resulting stress is a tensile stress; if the forces are reversed in direction, causing the bar to be compressed, they are called compressive stresses.

Ⅳ. Translate the following sentences into English.

1. 力的集度,即单位面积上的力,称为应力。
2. 应力常用希腊字母 σ 来表示。
3. 单位长度的伸长量称为应变,常用以下公式确定。
4. 材料力学是应用力学的一个分支。
5. 材料力学讨论固体在承受各种荷载时的性能。

Ⅴ. Answer the following questions briefly.

1. What is the concept of stress?
2. What is the concept of compression strain?
3. How do we obtain the tension strain?
4. What is the condition for Eq. 1-1 to be valid?
5. Please give us the units of the stress and strain, and show some examples.

Reading Materials

Statically Indeterminate Beams

In this paper we will consider the analysis of beams that have a larger number of *reactions*(反力) than the number of equations of static equilibrium. Such beams are said to be *statically indeterminate*(静不定的,超静定的), and their analysis requires that the deflections be taken into account. Only *statically determinate beams* (静定梁) were considered in the previous lesson, and in each instance we could immediately obtain the reactions of the beam by solving equations of static equilibrium. Knowing the reactions, we could then obtain the *bending moments*(弯矩) and *shear forces*(剪力), which in turn made it possible to find the stresses and deflections. However, when the beam is statically indeterminate, we cannot solve for the forces on the basis of statics alone. Instead we must take into account the deflection of the beam and obtain equations of compatibility to supplement the equations of statics. This same procedure was discussed for the case of statically indeterminate problems involving members in tension and compression.

Several types of statically indeterminate beam are illustrated in Fig. 1-2. The beam part(a) of the figure is fixed(or clamped) at support A and simply supported at B; such a beam is called either a *propped cantilever beam*(有支承悬臂梁) or a "fixed simple" beam. The reactions of the beam consist of horizontal and vertical forces at A, *a couple*(力偶) at A, and a vertical force at B. Because there are only three independent equations of static equilibrium for the beam, it is not possible to calculate all four of these reactions by statics. The number of reactions in excess of the number of equilibrium equations is called the *degree of statical indeterminacy* (超静定次数). Thus, the beam pictured in Fig. 1-2(a) is statical indeterminacy to the first degree. Any reactions in excess of the number needed to support the structure in a statically determinate manner are called *statical redundants*(赘余力), and the number of such redundants necessarily is the same as the degree of indeterminacy. For example, the reaction R_b shown in Fig. 1-2(a) may be considered as a redundant reaction. Note that when it is removed from the structure, there remains a cantilever beam. The statically determinate structure which remains when the redundant is released is called the *released structure*(放松结构) or the *primary structure*(基本结构). Another approach to the beam in Fig. 1-2(a) is to consider the reactive *moment*(弯矩、力矩)M_a as the redundant; if it is removed, the released structure is a simple beam with a *pin support*(铰支座) at A and a *roller support*(滚轴支座) at B.

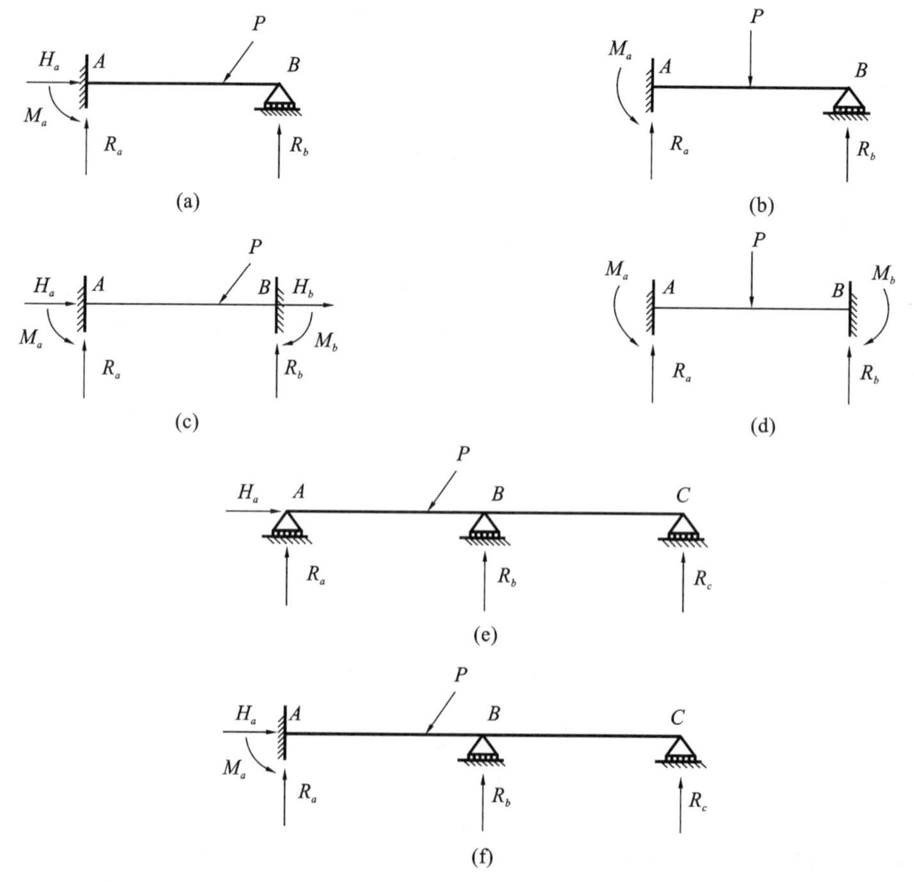

Fig. 1-2 Statically indeterminate beams

A special case arises if all loads on the beam are vertical (see Fig. 1-2(b)) because then the horizontal reaction vanishes. However, the beam is still statically indeterminate to the first degree inasmuch as there are now two independent equations of static equilibrium but three reactions.

A *fixed-end*(固定端) beam sometimes called a "fixed-fixed" beam, is shown in Fig. 1-2(c). At each support there are three reactive quantities, and hence the beam has a total of six unknown reactions. Because there are three *equations of equilibrium*(平衡公式), the beam is statically indeterminate to the third degree. If we take the reactions at one end as the three redundants, and if we remove them from the structure, a cantilever beam will remain as the released structure. If we remove the two fixed-end moments and one horizontal reaction, the released structure is a simple beam.

Again considering the special case of vertical loads only (see Fig. 1-2(d)), we find that now there are only four reactions to be determined. The number of static equilibrium equations is two and, therefore, the beam is statically indeterminate to the second degree.

The remaining two beams shown on Fig. 1-2 are examples of continuous beams, so called because they have more than one span and are continuous over a support. The beam shown in Fig. 1-2(e) is statically indeterminate to the first degree because there are four reactive forces and only three equations of static equilibrium. If R_b is selected as the redundant, and if we imagine it to be removed from the beam, then there will remain a statically determinate simple beam *AC*. If R_c is selected, the released structure will be a simple beam *AB* with an overhang *BC*. The last beam shown in the Fig. 1-2(f) is statically indeterminate to the second degree. We might select R_b and R_c as the redundant reactions, and then the released structure is a cantilever beam.

Reading Skills(1)—Reading without Translating

Some students complain that they can't read well in English. The primary reason, that they do not read well is that they simply do not read, they translate. They don't believe they understand anything, unless they translate it into Chinese. Reading and translating are not the same thing. Translating is an important skill. It can be helpful in the early stages of reading, but it should be left behind as soon as possible if you want to read more and faster.

Now read the following paragraphs without translating. Try to understand the meaning directly from English.

Both *uniaxial*(单轴的) and *biaxial*(双轴的) stresses are special cases of a more general stress condition known as *plane stress*(平面应力). An element *in plane*(平

面内）stress may have both *normal*（法向）and *shear*（剪切）stresses on the x and y faces, as shown in Fig. 1-3, but no stresses on the z face of the element. The shear stress on the x face of the element will be *denoted*（表示）by τ_{xy}, in which the first *subscript*（下标）denotes the plane on which the stress acts, and the second subscript denotes the direction of the shear stress.

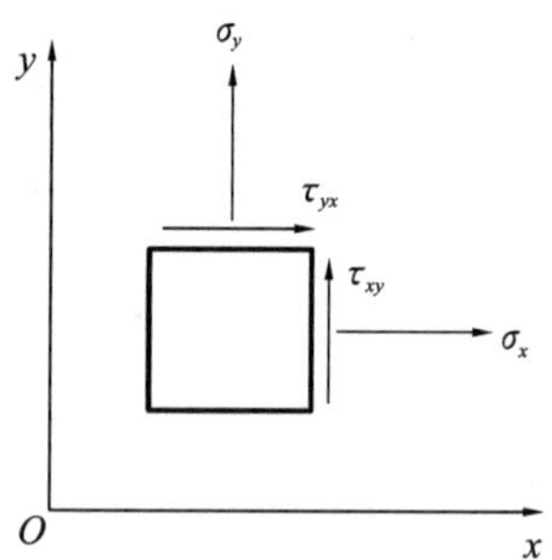

Fig. 1-3　Plane stress

When using this notating for identifying shear stresses, it is customary to assume that the shear stress is positive when it acts in the positive direction of the y axis. Thus, τ_{xy} is positive in the direction shown in the figure. Similarly, the shear stress on the upper face of the element is denoted by τ_{yx}, indicating that the stress acts on the y face of the element and is positive in the x direction.

Lesson 2

STRESSES IN COLUMNS

The average compressive stress in a centrally loaded column is founded by dividing the axial force by the cross-sectional area. The stress obtained in this manner for the case when the axial force is equal to the critical load is called the critical stress σ_{cr}. Thus, for the fundamental case of buckling, the critical load is given and the critical stress is(see Eq. 2-1)

$$\sigma_{cr}=\frac{P_{cr}}{A}=\frac{\pi^2 EI}{AL^2}=\frac{\pi^2 E}{(L/r)^2} \qquad 2\text{-}1$$

in which we have used the notation

$$r=\sqrt{\frac{I}{A}}$$

for the least radius of gyration of the cross section. The ratio L/r, which appears in the denominator of Eq. 2-1, is called the slenderness ratio of the column. We may note that the critical stress varies inversely with the square of this ratio. Of course, the stress σ_{cr}, must remain below the proportional limit of the material in order for Eq. 2-1 to be valid.

A diagram of the compressive stress in the column versus the slenderness ratio can now be plotted (see Fig. 2-1). The curve ABC is plotted from Eq. 2-1 and is called Euler's curve. This curve is valid physically only in the region BC where σ_{cr} is less than the proportional limit σ_{pl}. The limiting value of L/r above which Euler's column formula applies may be found by setting $\sigma_{cr}=\sigma_{pl}$ in Eq. 2-1 and solving for L/r. Taking structural steel as an example and assuming $\sigma_{pl}=36$ ksi and $E=30\ 000$ ksi, we find that the value of L/r corresponding to point B in Fig. 2-1 is approximately 91. Thus, for $L/r<91$ the average compressive stress in a simply supported ideal steel column will reach the proportional limit before buckling occurs; hence, the Euler formula for the critical load is inapplicable in this instance and gives values which are too high. If $L/r>91$, the column will fail by elastic buckling and Euler's formula may be used.

The curve BC in Fig. 2-1 shows that, when the slenderness ratio is large, the critical stress becomes very small; hence, a very slender column buckles at a low compressive stress. This condition cannot be improved by using a higher strength material; instead, the critical stress can be raised only by increasing the radius of gyration or by using a material with a higher modulus of elasticity.

When the slenderness ratio of a column is very low, we may expect the column to fail because of a failure of the material itself. Such a failure may take the form of crushing of the material, as occurs when the material is concrete; or yielding of the material, which occurs in the case of structural steel. ① Under these circumstances, a certain maximum compressive stress P/A can be established as the limit of the strength of the material, and the ultimate load can be determined accordingly. This limit is represented in Fig. 2-1 by the horizontal line DE drawn through the maximum stress σ_{max}; it represents a strength limit for the column.

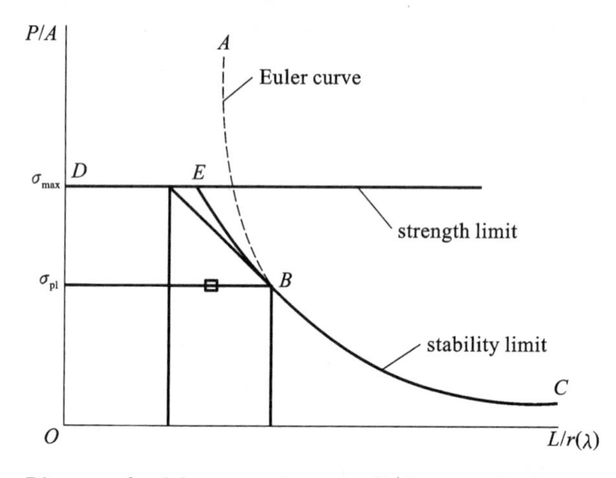

Fig. 2-1 Diagram of axial compressive stress P/A versus slenderness ratio L/r

Between the regions of short and long columns, there are a range of intermediate slenderness ratios too small for elastic stability to govern and too large for strength considerations alone to govern. ② Such medium-length columns buckle inelastically. For practical purposes it is sometimes sufficient to draw a straight line EB (see Fig. 2-1) and to consider that it represents the critical stresses for medium-length columns. In this manner we obtain the broken line $DEBC$ which can be used as the basis for designing columns of any length. Alternatively, a smooth curve connecting points D and B can be used.

The stress P/A obtained from a diagram of the kind shown in Fig. 2-1 must be considered as a maximum stress σ_{max} for the column. At this stress the column fails either by a direct failure of the material or by buckling, depending upon the slenderness ratio. The allowable working stress σ_w for compression must be taken as σ_{max}/n, where n is a factor of safety. The value chosen for n depends upon the probability of unforeseen or accidental increases in the load P, possible errors in the central applications of the load, and the likelihood of initial crookedness in the column. Imperfections in the column are apt to increase as the column becomes longer; therefore, it is logical to introduce a variable factor of safety which increases with the slenderness ratio. Typical values of the factor of safety for structural work are in the range 1.5 to 3.

Words and Expressions

column	[ˈkɔləm]	*n.* 柱
critical	[ˈkritikəl]	*adj.* 临界的
fundamental	[ˌfʌndəˈmentl]	*adj.* 基本的,原始的
		n. 基本原理,根本法则
buckle	[ˈbʌkl]	*v.* 压曲,纵向弯曲
gyration	[ˌdʒaiəˈreiʃən]	*n.* 回转,旋转
denominator	[diˈnɔmineitə]	*n.* 分母
slender	[ˈslendə]	*adj.* 细长的
inversely	[inˈvəsli]	*adv.* 相反地
versus	[ˈvəːsəs]	*prep.* ……对……(常略作 v 或 vs)
curve	[kəːv]	*n.* 曲线，曲线图表
strength	[streŋθ]	*n.* 强度
modulus	[ˈmɔdjuləs]	*n.* 模量,模数
crush	[krʌʃ]	*v.* 压碎,压坏
yielding	[ˈjiːldiŋ]	*adj.* 易弯曲的，柔顺的
govern	[ˈgʌvən]	*v.* 管理,支配,控制
failure	[ˈfeiljə]	*n.* 失败,故障,不足
probability	[ˌprɔbəˈbiləti]	*n.* 可能性,概率
unforeseen	[ˌʌnfɔːˈsiːn]	*adj.* 未预见到的,意料之外的
likelihood	[ˈlaiklihud]	*n.* 可能(性)
crooked	[ˈkrukid]	*adj.* 弯曲的,歪的
critical stress		临界应力
radius of gyration		回转半径
slenderness ratio		长细比
proportional limit		比例极限
Euler's (column) formula		欧拉(柱)公式
ksi=kilopounds per square inch		千磅每平方英寸
strength limit		强度极限
axial compressive		轴向受压的
medium-length column		中长柱
broken line		折线
allowable (working) stress		许可(工作)应力
be apt to do		易于……,有倾向于……

Notes

① Such a failure may take the form of crushing of the material, as occurs when the material is concrete; or yielding of the material, which occurs in the case of structural steel.

其中的 such…as 在不同的结构中有不同的解释。such…as+名词时,as 相当于介词 like; such…as+从句时,as 一般可作 which(that)、who(whom)、when 解释,引导的是定语从句。全句意思是:如材料为混凝土,这种破坏可以是材料压碎的形式;如材料为钢材,这种破坏则是使材料弯曲的形式。

② Between the regions of short and long columns, there are a range of intermediate slenderness ratios too small for elastic stability to govern and too large for strength considerations alone to govern.

句中连续出现两个 too…to 结构,其中的 for elastic stability 和 for strength considerations 分别是两个不定式 to govern 的逻辑主语。全句意思是:在短柱和长柱范围之间,有一段属于中间长细比,其用弹性稳定控制太小,而单独用强度控制又太大。

Exercises

Ⅰ. Translate the following words into Chinese.

1. the critical load ____________________
2. the ultimate load ____________________
3. a modulus of elasticity ____________________
4. a factor of safety ____________________
5. elastic buckling ____________________
6. initial crookedness ____________________
7. the slenderness ratio ____________________

Ⅱ. Translate the following words into English.

1. 临界应力____________________
2. 欧拉公式____________________
3. 比例极限____________________
4. 强度极限____________________
5. 非弹性屈曲____________________
6. 平均压应力____________________
7. 回转半径____________________
8. 许可(工作)应力____________________

Ⅲ. Translate the following sentences into Chinese.

1. A diagram of the compressive stress in the column versus the slenderness ratio can now be plotted.

2. For practical purposes it is sometimes sufficient to draw a straight line *EB* and to consider that it represents the critical stresses for medium-length columns.
3. At this stress the column fails either by a direct failure of the material or by buckling, depending upon the slenderness ratio.
4. This condition cannot be improved by using a higher strength material; instead, the critical stress can be raised only by increasing the radius of gyration or by using a material with a higher modulus of elasticity.
5. The Euler formula for the critical load is inapplicable in this instance and gives values which are too high.

Ⅳ. Translate the following sentences into English.

1. 我们注意到临界应力的变化与柱子长细比的平方成反比。
2. 如果 $L/r>91$,柱子将由于弹性屈曲而失效,欧拉公式可以应用。
3. 曲线 *ABC* 是根据公式 2-1 绘制的,被称为欧拉曲线。
4. 结构工程安全系数的典型值为 1.5～3。

Ⅴ. Answer the following questions briefly.

1. What is called the critical stress?
2. What is called the slenderness ratio of the column?
3. What does the curve *BC* in Fig. 2-1 show?
4. What will happen when the slenderness ratio is very low?
5. Is it true that "the column buckling, when its slenderness is large, can be improved by using a higher strength material"? What is the reason?

Reading Materials

Required Strengths of Concrete

The different strengths of a particular concrete may be required in the design of reinforced concrete structures; these include the compressive, tensile, *flexural*(弯曲的), and *shear strengths*(剪切强度). Although it is sometimes difficult to perform tests to predict these strengths, it is possible to relate them to the compressive strength.

In designing structural members, it is assumed that the concrete resists compressive stresses and not tensile stresses; therefore, compressive strength is the *criterion*(标准、准则) of quality of concrete. The other concrete stresses can be taken as a percentage of the compressive strength, which can be easily and accurately determined from tests. Specimens used to determine compressive strength may be cylindrical, cubical, or prismatic.

Test specimens(试块、样本) in the form of a cylinder 6 in. (150 mm) across and 12 in. (300 mm) high are usually used to determine the strength of concrete. The failure of the concrete specimen can be in one of three modes, as shown in Fig. 2-2.

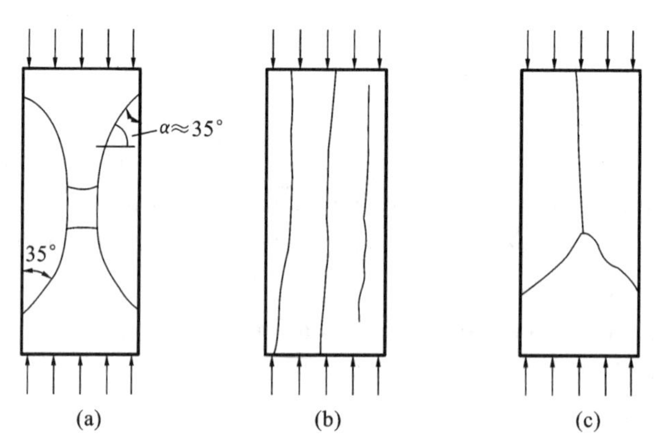

Fig. 2-2 Failure of the concrete specimen

First, under axial compression, the concrete specimen may fail in shear. Resistance to failure is due to both cohesion and internal friction. The shear resistance may be represented by Coulomb's equation and Mohr's circle, as shown in Fig. 2-3.

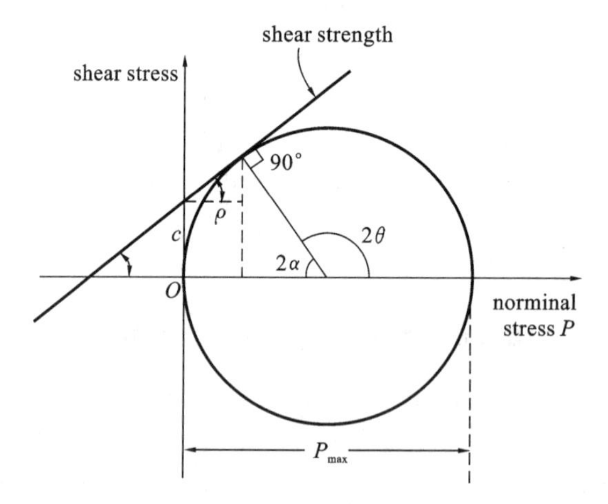

Fig. 2-3 Mohr's circle

The second type of failure (see Fig. 2-2(b)) results in the separation of the specimen into column or piece by what is known as splitting or fracture. This failure occurs when the strength of concrete is high, and lateral expansion at the end bearing surfaces is relatively unrestrained.

The third type of failure (see Fig. 2-2(c)) is seen when a combination of shear and splitting failure occurs.

Fig. 2-4 shows typical stress-strain curves for concretes of different strengths. All curves consist of an initial relatively straight elastic portion, reaching maximum stress at a strain of about 0.002; then rupture occurs at a strain of about 0.003.

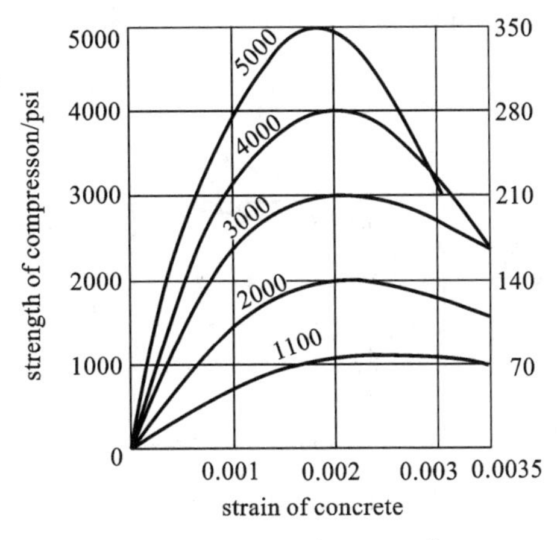

Fig. 2-4 Typical stress-strain curves for concretes

If a concrete test specimen is loaded over a part of its area, the strength is increased owing to the retaining influence of the surrounding concrete of the unloaded part. The ultimate strength in local compression or crushing may be found from the following Eq. 2-2

$$N_l \leqslant \gamma A_l f \qquad 2\text{-}2$$

Where

$\gamma = 1 + 0.35\sqrt{\frac{A_0}{A_l} - 1}$;

f = ultimate strength of cylinder or prism fully loaded;

A_0 = total area of specimen;

A_l = loaded area.

The area ratio $\gamma \leqslant 1.5$ for the action of a local load only, $\gamma \leqslant 2.0$ for the action of local and other loads.

Direct tension tests are not reliable for predicting the tensile strength of concrete, due to minor *misalignment*(不对齐) and stress concentrations in the gripping devices. An indirect tension test in the form of splitting a 6 by 12 in. (150 by 300 mm) cylinder was suggested by the Brazilian Fernando Carneiro (see Fig. 2-5). The test is usually called the *splitting test*(劈裂试验), but it was also named the Brazilian test.

Pure shear is seldom encountered in reinforced concrete members, as it is usually accompanied by the action of normal forces. An element subjected to pure shear breaks *transversely*(横向地) into two parts. Therefore, the concrete element must be strong enough to resist the applied shear forces.

Shear strength may be considered as 20 to 30 percent greater than the tensile strength of concrete, or about 12 percent of its compressive strength. The American Concrete Institute (ACI，美国混凝土协会) code allows an ultimate shear strength of 2 psi (0.17 N/mm²) on *plain concrete* (素混凝土) sections.

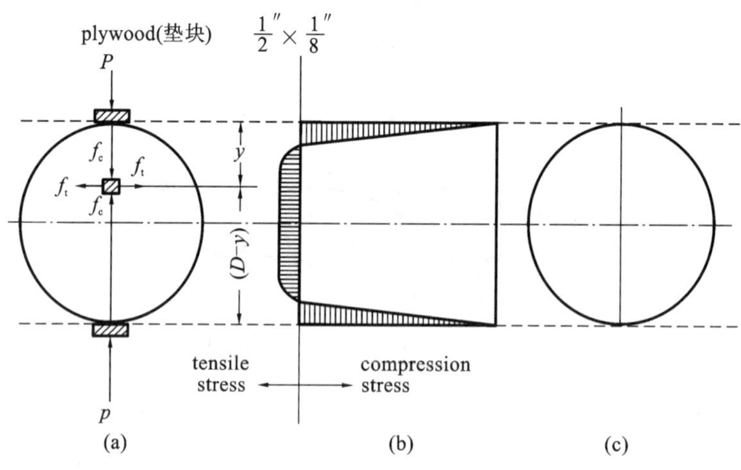

Fig. 2-5 Splitting test

(a) splitting test of cylinder; (b) distribution of horizontal stress; (c) cylinder after test

Reading Skills(2)—Silent Reading

Reading is essentially silent. The students should read without sound, without even lip movement because the eyes are much quicker than the lips.

Try to read the material given below silently without even lip movement.

In our previous discussions, we used a sign convention for the shear stress τ_v which was based upon whether the shear stress acted *clockwise*(顺时针方向转动地) or *counterclockwise*(逆时针方向转动地) against the element. We will continue to use that convention for τ_θ in our consideration of plane stress, therefore, we see that on the plane for which $\theta=0°$, we have $\tau_\theta=-\tau_{xy}$, and on the plane $\theta=90°$, we have $\tau_\theta=\tau_{yx}$. Also, from the equality of shear stresses on perpendicular planes, it is obvious that

$$\tau_{xy}=\tau_{yx}$$

Reading Skills(3)—Reading by Phrases

The students should practise not to read *word by word*(逐字地) but to read groups of words. Take the following sentence for example:

Being able to read by phrases instead of by single words results from practice.

If you read word by word, your eyes must stop 14 times. When you read two or three words at a single glance, your eyes may stop six times only.

Now try to read the above reading materials not by individual words, but by phrases.

Reading Skills(4)—Scanning

Like skimming, scanning is also quick reading. However, in this case the search is more focused. To scan is to read quickly in order to locate specific information. When you read to find a particular date, name, or number you are scanning. Because scanning is directed and purposeful, it should be very fast. As you scan your eyes over a page or down a column, keep in mind the specific information you are looking for. Do not read closely. The obvious way to scan is to run your eyes, rapidly over the words or figures, neglecting all others but what you are looking for.

In the following paragraphs, you should scan to find specific information. You should find each answer in less than one minute.

Questions: (1) What factors determine the use of a material?

The specifications of the designers have to be matched against what is known about a material's strength, how easily it conducts electricity, how quickly it corrodes, etc. But the material chosen for a given application is the one which must cheaply meet the specifications of the designers. The two dominant factors which determine the use of a material are its cost and its physical and chemical properties.

(2) What's the characteristic of supports for simply supported beam?

The beam in Fig. 2-6, with a pin support at one end a roller support at the other, is called a simply supported beam, or a simple beam. The essential feature of a simply supported beam is that both ends of the beam may rotate freely during bending, but they cannot translate in the *lateral*(横向的,侧向的) direction (that is, *transverse*(横向的,横断的) to the axis. Also, one end of the beam can move freely in the axial direction (that is, horizontally). The supports of a simple beam may sustain vertical reactions acting either upward or downward.

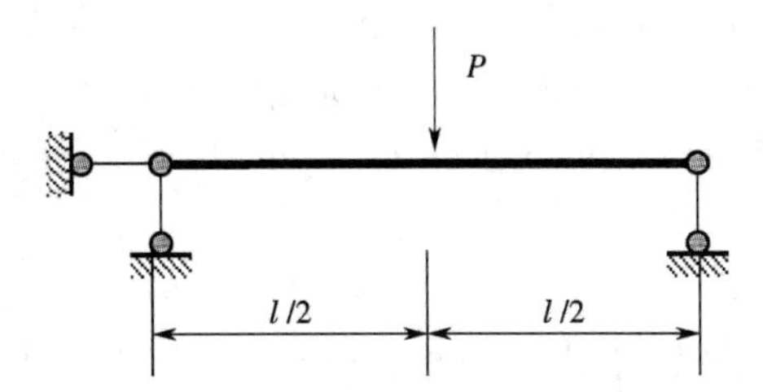

Fig. 2-6 A simple beam under concentrated force

Lesson 3

STRUCTURAL SAFETY AND RELIABILITY

Safety and serviceability

To serve its purpose, a structure must be safe against collapse and serviceable in use. Serviceability requires that deflections be adequately small, that cracks, if any, be kept to tolerable limits, that vibrations be minimized, etc. ① Safety requires that the strength of the structure be adequate for all loads which may foreseeably act on it. If the strength of a structure, built as designed, could be predicted with precision, and if the loads and their internal effects (moments, shears, axial forces) were known with equal precision, safety could be assured by providing a carrying capacity just barely in excess of the known loads. However, there are a number of sources of uncertainty in the analysis, design, and construction of reinforced concrete structures. These sources of uncertainty, which require a definite margin of safety, may be listed as follows.

①Actual loads may differ from those assumed in the design.

②Actual loads may be distributed in a manner different from that assumed in the design.

③The assumptions and simplifications inherent in any analysis may result in calculated load effects-moments, shears, etc. —different from those which in fact act in the structure.

④The actual structural behavior may differ from that assumed, owing to imperfect knowledge.

⑤Actual member dimensions may differ from those specified by the designer.

⑥Reinforcement may not be in its proper position.

⑦Actual material strength may be different from that specified by the designer.

In addition, in establishing a safety specification, consideration must be given to the consequences of failure. In some cases, a failure would merely be an inconvenience. In other cases, loss of life and significant loss of property may be involved. A further consideration should be the nature of the failure, should it occur. ② A gradual failure with ample warning permitting remedial measures is preferable to a sudden, unexpected collapse.

It is evident that the selection of an appropriate margin of safety is not a simple matter. However, progress has been made toward rational safety provisions in design codes[3].

Loads

The loads which act on structures can be divided into three broad categories: dead loads, live loads, and environmental loads.

Dead loads are those which are constant in magnitude and fixed in location throughout the lifetime of the structure. Usually the major part of the dead load is the weight of the structure itself. Dead loads can generally be calculated with good precision from the design configuration and dimensions of the structure.

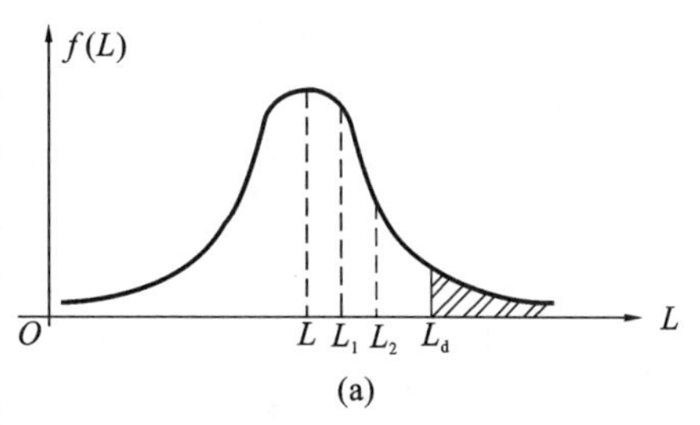

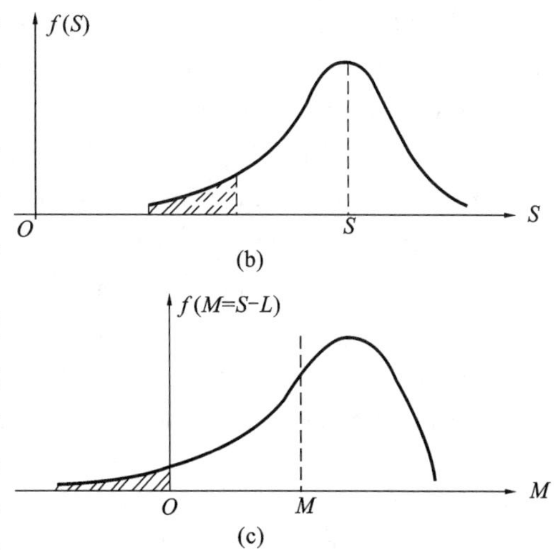

Fig. 3-1 Frequency curves for load L, strength S, and safety margin M

(a) load L; (b) strength S; (c) safety margin $M = S - L$

Live loads consist chiefly of occupancy loads in buildings and traffic loads on bridges. They may be either fully or partially in place or not present at all and may also change in location. Their magnitude and distribution at any given time are uncertain, and even their maximum intensities throughout the lifetime of the structure are not known with precision.

Environmental loads consist mainly of snow loads, wind pressure and suction, earthquake actions (i. e., inertia forces caused by earthquake motions), soil pressures on subsurface portions of structures, loads from possible ponding of rainwater on flat roofs, and forces by temperature differentials. Like live loads, environmental loads at any given time are uncertain in both magnitude and distribution. Specified values do not constitute average values but represent expected upper limits.

Since the maximum load which will occur during the life of a structure is uncertain, it can be considered a random variable. In spite of this uncertainty, the engineer must provide an adequate structure. A probability model for the maximum load can be devised by means of a probability density function for loads, as represented by the frequency curve of Fig. 3-1(a). The exact form of this distribution curve, for any particular type of loading such as office loads, can be determined only on the basis of statistical data obtained from large-scale load surveys. A number of such surveys have been carried out or are in progress. For types of loads for which such data are scarce, fairly reliable information can be obtained from experience observation, and judgment.

In such a frequency curve, Fig. 3-1(a), the area under the curve between two abscissas, such as the load L_1 and load L_2, represents the probability of occurrence of load L of magnitude $L_1 < L < L_2$. A specified service load L_d for design is selected conservatively in the upper region of L in the distribution curve, as shown. The probability occurrence of loads larger than L_d is then given by the shaded area to the right of L_d. [4] It is seen that the specified service load is considerably larger than the mean load L acting on the structure. This mean load is much more typical of average load conditions than the design load L_d and is sometimes known as the characteristic load.

Strengths

The strength of a structure depends primarily on the strength of the materials from which it is made. For this purpose, minimum material strengths are specified in standardized ways. Actual material strengths cannot be precisely known and therefore also constitute random variables. Structural strength depends, furthermore, on the care with which a structure is built, which in turn reflects the quality of supervision and inspection. Member sizes may differ from specified dimensions, reinforcement may be out of position, poorly placed concrete may show voids, etc.

Strength of the entire structure or of a population of repetitive structures, e. g., highway overpasses, can also be considered a random variable with probability density function of the type shown in Fig. 3-1(b). As in the case of loads, the exact form of this function cannot be known but can be approximated as known date, such as statistics of actual, measured materials and member strengths and similar information. Considerable information in this type has been, or is being, developed from utilized.

Structural safety

A given structure has a safety margin M if

$$M = S - L > 0 \qquad 3\text{-}1$$

i. e., if the strength of the structure is larger than the load acting on it. Since S and L are random variables, the safety margin $M = S - L$ is also a random variable. A plot of the probability function of M may appear as in Fig. 3-1(c), failure occurs when M is less than 0. Thus, the probability of failure is represented by the shaded area in the figure.

Even though the precise form of the probability density function for S and L and therefore for M, is not known, much can be achieved in the way of a rational approach to structural safety. One such approach is to require that the mean safety margin $\overline{M}$ be a specific number β of standard deviations σ_m above 0. It can be demonstrated that this results in the requirement that

$$\gamma_S \overline{S} \geqslant \gamma_L \overline{L} \quad 3\text{-}2$$

where γ_L is a partial safety coefficient larger than L applied to the mean load $\overline{L}$ and γ_S is a partial safety coefficient smaller than S applied to the mean strength $\overline{S}$. The magnitude of each partial safety coefficient depends on the variance of the quantity to which it applies, L and S, and on the chosen value of β, the safety index of the structure. As a general guide, a value of the safety index β between 3 and 4 corresponds to a probability of failure of the order of 1 : 1 000 000. In practice, the value of β is often established by calibration against well proved and established design.

Words and Expressions

reliability	[riˌlaiəˈbiləti]	*n.* 可靠性
serviceability	[ˌsəːvisəˈbiləti]	*n.* 适用性
collapse	[kəˈlæps]	*n.* 倒塌,垮掉
deflection	[diˈflekʃən]	*n.* 挠曲,挠度,偏移,偏差
tolerable	[ˈtɔlərəbl]	*adj.* 可允许的,可容忍的
precision	[priˈsiʒən]	*n.* 精确,精密度,精度
inherent	[inˈhiərənt]	*adj.* 固有的,内在的
reinforcement	[ˌriːinˈfɔːsmənt]	*n.* 加强,加固
specification	[ˌspesifiˈkeiʃən]	*n.* 规程,规范,说明书
remedial	[riˈmiːdjəl]	*adj.* 补救的,纠正的
rational	[ˈræʃənəl]	*adj.* 合理的
provision	[prəuˈviʒən]	*n.* 条款,设备,供应
configuration	[kənˌfigjuˈreiʃən]	*n.* 外形,构造
suction	[ˈsʌkʃən]	*n.* 吸入
subsurface	[ˈsʌbˈsəːfis]	*adj.* 地表下的
		n. 地表下的岩石或土壤
pond	[ˈpɔnd]	*n.* 池塘
		v. 拦截,堵塞
random	[ˈrændəm]	*adj.* 随机的
devise	[diˈvaiz]	*v.* 设计,想出
abscissa	[æbˈsisə]	*n.* 横坐标
calibration	[ˌkæliˈbreiʃən]	*n.* 校准,刻度
to serve its purpose		为了达到其目的
axial force		轴力
carrying capacity		承载能力

reinforced concrete	钢筋混凝土
margin of safety	安全储备,安全系数,可靠度
design codes	设计规范
environmental loads	环境荷载
wind pressure and suction	风压力和吸力
random variable	随机变量
by means of	通过,用,以
a probability density function for load	荷载的概率密度函数
in progress	在进行中
mean load	平均荷载
characteristic load	特征荷载(荷载特征值)
probability of failure	失效概率
standard deviations	标准差
partial safety coefficient	分项安全系数

Notes

① Serviceability requires that deflections be adequately small, that cracks, if any, be kept to tolerable limits, that vibrations be minimized, etc.
其中 requires 后面跟随三个宾语从句,都用虚拟结构。句中 if any 是一个省略的插入句,可理解为 if there are any cracks。

② A further consideration should be the nature of the failure, should it occur.
...should it occur 是以倒装表示虚拟的结构,即 should 移至主语 it 的前面。

③ 在英语中,规范一般用 code 来表示,而规程、规定一般用 specification 来表示。规范、规程中的条款可用 provision 或 specification。

④ The probability occurrence of loads larger than L_d is then given by the shaded area to the right of L_d.
The probability occurrence of loads larger than L_d是形容词(比较级)短语作后置定语。全句意思是:荷载大于 L_d 出现的概率为 L_d 右边阴影部分的面积。

Exercises

Ⅰ. Translate the following words into Chinese.

1. dead loads ______
2. live loads ______
3. the mean load ______
4. rational safety provisions ______
5. the characteristic load ______
6. random variable ______

Ⅱ. Translate the following words into English.

1. 温度差________________
2. 补救措施________________
3. 安全度规范________________
4. 雨水积聚________________
5. 概率密度函数________________
6. 概率曲线________________
7. 失效概率________________
8. 分项安全系数________________

Ⅲ. Translate the following sentences into Chinese.

1. Safety requires that the strength of the structure be adequate for all loads which may foreseeably act on it.
2. Serviceability requires that deflections be adequately small, that cracks, if any, be kept to tolerable limits, that vibrations be minimized, etc.
3. Actual loads may differ from those assumed in the design.
4. Actual loads may be distributed in a manner different from that assumed in the design.
5. Dead loads are those which are constant in magnitude and fixed in location throughout the lifetime of the structure.

Ⅳ. Translate the following sentences into English.

1. 结构必须具备抗破坏的安全度和使用时的适用性。
2. 作用在结构上的荷载可分为三大类。
3. 像活荷载一样，某时刻的环境荷载无论是大小还是分布都是不确定的。
4. 规定值并不是平均值，而是表示期望的上限。
5. 设计规范中合理的安全条款已经取得了很大进展。
6. 由此可看出，这个规定的使用荷载比作用在结构上的平均荷载大得多。

Ⅴ. Answer the following questions briefly.

1. Is it true that "Failure will never occur in a well-designed structure"?
2. What does serviceability require?
3. What does safety require?
4. What consideration must be given in establishing a safety specification?

Reading Materials

Load Combinations

The safety provisions of the ACI code are given in the form of Eq. 3-2 using strength reduction factors and load factors. These factors are based to some extent on statistical information but to a larger degree on experience, engineering judgment, and compromise. In words, the design strength S_n of a structure or member must be at least equal to the required strength U calculated from the factored loads, i. e. ,

$$\text{design strength} \geqslant \text{required strength}$$

or

$$\phi S_n \geqslant U \qquad 3\text{-}3$$

The nominal strength S_n is computed (usually somewhat conservatively) by accepted methods. The required strength U is calculated by applying appropriate load factors to the respective service loads: dead load D, live load L, wind load W, earthquake load E, earth pressure H, fluid pressure F, impact allowance I, and environmental effects T that may include *settlement*(沉降), *creep*(徐变), shrinkage, and temperature change. Loads are defined in a general sense, to include either loads or the related internal effects such as moments, shears, and *thrusts*(推力). Thus, in specific terms for a member subjected, say, to moment, shear, and axial load:

$$\phi M_n > M_u \qquad 3\text{-}4(a)$$

$$\phi V_n > V_u \qquad 3\text{-}4(b)$$

$$\phi P_n > P_u \qquad 3\text{-}4(c)$$

where the subscripts n denote the nominal strengths in *flexure*(弯曲), shear, and axial load, respectively, and the subscripts u denote the factored load moment, shear, and axial load. In computing the factored load effects on the right, load factors may be applied either to the service loads themselves or to the internal load effects calculated from the service loads.

The load factors specified in the ACI code, to be applied to calculated dead loads and those live and environmental loads specified in the appropriate codes or standards, are summarized in Table 3-1. These are consistent with the concepts introduced in SEI/ASCE 7, *Minimum Design Loads for Buildings and Other Structures*, and allow design of composite structures using combinations of structural steel and reinforced concrete. For individual loads, lower factors are used for loads known with greater certainty, e. g. , dead load, compared with loads of greater variability, e. g. , live loads.

Table 3-1 Factored load combinations for determining required strength U in the ACI code

Condition	Factored load or load effect U
Basic	$U=1.2D+1.6L$
Dead plus fluid	$U=1.4(D+F)$

续表

Condition	Factored load or load effect U
Snow, rain, temperature, and wind	$U=1.2(D+F+T)+1.6(L+H)+0.5(L_r \text{ or } S \text{ or } R)$
	$U=1.2D+1.6(L_r \text{ or } S \text{ or } R)+(1.0L \text{ or } 0.8W)$
	$U=1.2D+1.6W+1.0L+0.5(L_r \text{ or } S \text{ or } R)$
	$U=0.9D+1.6W+1.6H$
Earthquake	$U=1.2D+1.0E+1.0L+0.2S$
	$U=0.9D+1.0E+1.6H$

Note: L_r = roof live load; R = rain load; S = snow load.

Further, for load combinations such as dead plus live loads plus wind forces, *reductions*(折减) are applied to one load or the other that reflect the improbability that an excessively large live load coincides with an unusually high windstorm.

The factors also reflect, in a general way, uncertainties with which internal load effects are calculated from external loads in systems as complex as highly indeterminate, inelastic reinforced concrete structures which, in addition, consist of variable-section members (because of tension cracking, discontinuous reinforcement, etc.).

Finally, the load factors also distinguish between two situations, particularly when horizontal forces are present in addition to gravity, i. e., the situation where the effects of all simultaneous loads are *additive*(相叠加的), as distinct from that in which various load effects *counteract*(抵消的)each other. For example, in a retaining wall the soil pressure produces an *overturning*(倾覆)moment, and the gravity forces produce a counteracting stabilizing moment.

In all cases in Table 3-1, the controlling equation is the one that gives the largest factored load effect U.

The strength reduction factors ϕ in the ACI code are given different values depending on the state of knowledge, i. e., the accuracy with which various strengths can be calculated. Thus, the value for bending is higher than that for shear or bearing. Also, ϕ values reflect the probable importance, for the survival of the structure, of the particular member and of the probable quality control achievable. For both these reasons, a lower value is used for columns than for beams. Table 3-2 gives the ϕ values specified in the ACI code.

The joint application of strength reduction factors (see Table 3-2) and load factors (see Table 3-1) is aimed at producing approximate probabilities of under strength of the order of 1/100 and of overloads of 1/1 000. This results in a probability of structural failure of the order of 1/100 000.

Table 3-2 Strength reduction factors in the ACI code

Strength conduction		Strength reduction factors ϕ
Tension-controlled section		0.9
Compression with spiral reinforcement	Members with spiral reinforcement	0.70
	Other reinforced members	0.65
Shear and torsion		0.75
Bearing on concrete		0.65
Post-tensioned anchorage zones		0.85
Strut-and-tie models		0.75

In addition to the values given in Table 3-2, ACI code Appendix B, "Alternative Provisions for Reinforced and Prestressed Concrete Flexural and Compression Member", allows the use of load factors and strength reduction factors from previous editions of the ACI code. The load factors and strength reduction factors of ACI code Appendix B are calibrated in conjunction with the detailed requirements of that appendix. Consequently, they may not be interchanged with the provisions of the main body of the code.

Reading Skills(5)—Critical Reading

Critical(批判性的) reading demands that readers make judgments about what they read. This kind of reading requires posing and answering questions such as:

Does my own experience support that of the author?

Do I share the author's point of view?

Am I *convinced*(深信) of the author's arguments and evidence?

Now read the following paragraphs with some questions.

In contrast with many other highly developed countries, the United States does not have an official, national code governing structural concrete. The responsibility for producing and maintaining design *specifications*(规程，规定) rests with various professional groups, trade associations, and technical *institutes*(协会) that have produced the needed documents.

The American Concrete Institute (ACI) has long been a leader in such efforts. As one part of its activity, the American Concrete Institute has published the widely recognized *Building Code Requirements for Structural Concrete*, which serves as a guide in the design and construction of reinforced concrete buildings. The ACI code has no official status in itself. However, it is generally regarded as an authoritative statement of current good practice in the field of reinforced concrete. As a result, it has been incorporated by law into countless *municipal*(地方性的) and regional building codes that do have legal status. Its provisions thereby attain, in effect, legal standing. Most reinforced concrete buildings and related construction in the United States are designed in accordance with the current ACI code. It has also served as a model document for many other countries. A second ACI publication, *Commentary on Building Code Requirements for Structural Concrete*, provides background material and rationale for the code provisions. The American Concrete Institute also publishes important journals and standards, as well as recommendations for the analysis and design of special types of concrete structures such as the tanks.

Most highway bridges in the United States are designed according to the requirements of the America Association of State Highway and Transportation Officials(AASHTO) bridge specifications which not only contain the provisions relating to loads and load distributions mentioned earlier, but also include detailed provisions for the design and construction of concrete bridges. Many of the provisions follow ACI code provisions closely, although differences will be found.

The design of railway bridges is done according to the specifications of the AREA Manual of Railway Engineering. It, too, is patterned after the ACI code in most respects, but it contains most additional material pertaining to railway structures of all types.

No code or design specification can be *construed*(解释,分析) as a substitute for sound engineering judgment in the design of concrete structures. In structure practice, special *circumstances*(详情) are frequently encountered where code provisions can serve only as a guide, and the engineer must rely upon a firm understanding of the basic principles of structural mechanics applied to reinforced or prestressed concrete, and an intimate knowledge of the nature of the materials.

Reading Skills(6)—Main Idea

Being able to determine the main idea of a passage is one of the most useful reading skills you can develop. It is a skill you can apply to any kind of reading.

The main idea of a passage is the most important message presented by the author; it is the thought which is presented from the beginning to the end of the passage; in a well-written paragraph, most of the sentences support, describe, or explain the main idea of that paragraph. The main idea is sometimes stated in the first or last sentence of the paragraph. Sometimes it is only implied.

In order to determine the main idea of a paragraph, you should ask yourself what idea is common to most of the sentences in the paragraph; what is the idea that relates these sentences to each other? What opinion do they all support?

Now read the text again, then read the following material. Try to find the main idea of each paragraph.

In this paper reinforced concrete (in-situ and precast) and prestressed concrete are considered.

In the mid-1982s an internationally valid state-of-the-art review for resistance of concrete was exceptionally difficult to write, partly because of the *enormous*(巨大的) amount of literature on the subject and partly because of the large differences in attitude in different countries towards certain aspects of design. It is to be hoped that the outcome of some research programs on beam-column joints will greatly reduce the latter source of *divergence*(分歧). In the meantime it is hoped that the following discussion is reasonably balanced.

Reinforcement controls and delays failure in concrete members, the degradation process generally being initiated by cracking prevents in the concrete. Inelastic *elongation*(伸长) of reinforcement within a crack prevents the latter from closing when the load direction is reversed and *cyclic*(循环的) loading leads to progressive crack widening and steel yielding. Fenwick argued that shear in plastic hinge regions of beams is resisted by truss action until the phase of rapid strength degradation in which large shear displacements occur.

Lesson 4

WIND LOAD ON BUILDINGS

When the undisturbed air flow approaches a building, it is forced around and over the building. This creates areas of pressure or suction on the building facades, gables and roofs. Pressure and suction refer to air pressures above and below barometric pressure levels, respectively. Pressure is marked by plus and suction is marked by minus in the following figures.

Wind perpendicular to building facades

Wind loads on roofs depend on the roof shape. On a pitched roof with a slope exceeding approximately 35°, there will be pressure on the windward part of the roof and suction on the leeward part of the roof(see Fig. 4-1, 4-2). For flat roofs and for pitched roofs with a slope of less than 15°, suction affects the whole roof.

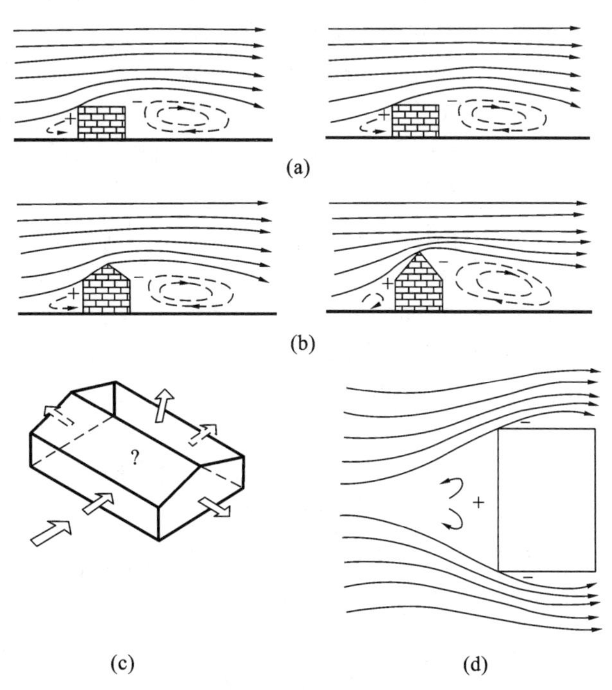

Fig. 4-1 Wind load on external walls and pitched roof. The direction of the wind is perpendicular to the longitudinal direction of the building

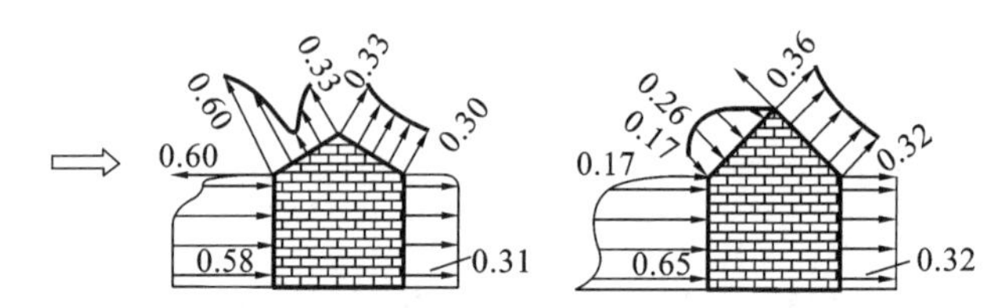

Fig. 4-2 Distribution of pressure and suction on pitched roofs with 30° and 45° inclination, respectively

For pitched roofs with a slope of between 15° and 30°, suction as well as pressure may occur on the upwind part of the roof. Euro code Ⅰ includes this by giving two load cases, one specifying suction and one specifying pressure on the upwind part of the roof.

Canopy roofs are subject to vertical, lifting forces below the windward canopy. This load acts with the wind load on the upper roof surfaces.

Wind parallel with building facades

When the wind flows along building facades, suction will always affect the roof. The greatest suction is obtained close to the upwind gable end.

The suction on pitched roofs is much greater than the suction on hipped roofs. In Fig. 4-3, the hipped roof has the same roof inclination at facades and gable ends. The wind direction is along the building and the pressure coefficients shown refer to suction on roof areas of 10 m^2 at the upwind roof corner. This effect is very pronounced when considering roof failures in hurricane areas. Failures are often prevented on structures with hipped roofs, whereas many flat roofs and pitched roofs are blown off during strong hurricanes. ① If this aspect was incorporated into the building traditions of areas prone to severe hurricanes, the number of low-rise structural failures in these areas could be reduced in the future.

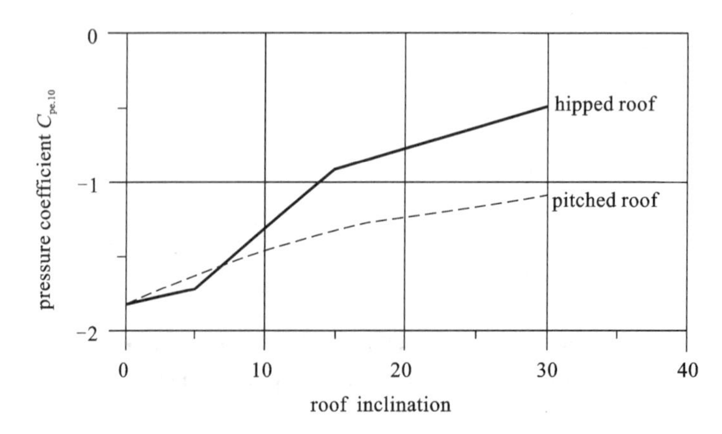

Fig. 4-3 Suction on pitched roofs and hipped roofs, respectively, based on Euro code Ⅰ

Words and Expressions

suction	[ˈsʌkʃən]	*n.* 吸,吸入
facade	[fəˈsɑːd]	*n.* 正面
gable	[ˈgeibl]	*n.* 山墙,尖顶屋两端的山形墙
barometric	[ˌbærəˈmetrik]	*adj.* 气压(表)的
windward	[ˈwindwəd]	*adv.* 向风,迎风,上风
		adj. 上风的
		n. 上风面
leeward	[ˈliːwəd, ˈluːəd]	*adj.* 背风的
		adv. 向下风
		n. 下风
longitudinal	[ˌlɔndʒiˈtjuːdinəl]	*adj.* 纵向的,经度的
upwind	[ˈʌpˈwind]	*n.* 逆风
		adj. 逆风的
canopy	[ˈkænəpi]	*n.* 天篷,顶篷
vertical	[ˈvəːtikəl]	*adj.* 垂直的
		n. 垂直线
lift	[lift]	*v.* 举起
pronounced	[prəˈnaunst]	*adj.* 显著的,断然的,明确的
hurricane	[ˈhʌrikən]	*n.* 飓风
pitched roof		(双坡)斜屋顶
Euro code I		欧洲规范 I
(be) subject to		服从,以……为条件,受……的支配;易遭受(发生)
hipped roof		四坡屋顶
prone to		倾向于……
low-rise		低层的

Notes

① Failures are often prevented on structures with hipped roofs, whereas many flat roofs and pitched roofs are blown off during strong hurricanes.
四坡屋顶的结构常可以避免被破坏，然而在强飓风中许多平屋顶和双坡屋顶经常被吹走。

Exercises

Ⅰ. Translate the following words into Chinese.

1. flat roof ______
2. pitched roof ______
3. upwind part of the roof ______
4. canopy roof ______
5. windward part of the roof ______
6. wind parallel with building facades ______
7. wind perpendicular to building facades ______

Ⅱ. Translate the following words into English.

1. 风吸力______
2. 风压力______
3. 四坡屋顶______
4. 建筑正立面______
5. 倾角为30°的坡屋顶______
6. 低层建筑______

Ⅲ. Translate the following sentences into Chinese.

1. When the undisturbed air flow approaches a building, it is forced around and over the building.
2. Pressure is marked by plus and suction is marked by minus in the following figures.
3. For pitched roofs with a slope of between 15°and 30°, suction as well as pressure may occur on the upwind part of the roof.
4. If this aspect was incorporated into the building traditions of areas prone to severe hurricanes, the number of low-rise structural failures in these areas could be reduced in the future.

Ⅳ. Translate the following sentences into English.

1. 风压力和吸力分别指的是空气的压力大于或者小于大气压力。
2. 当屋顶的坡度超过35°时,在屋顶的迎风面为压力,在屋顶的背风面为吸力。
3. 雨棚下部受垂直升力的作用,该荷载与上部风荷载同时作用其上。

Ⅴ. Answer the following questions briefly.

1. Is it true that "The suction on hipped roofs is much greater than the suction on pitched roofs"?
2. Is it true that "For pitched roofs suction will occur on the upwind part of the roof"?

Reading Materials

Static Wind Load According to Euro Code Ⅰ

The static wind load specified in Euro code Ⅰ is divided into wind pressures acting on surfaces and global wind forces.

Wind pressures

Wind pressures acting on external and internal surfaces, respectively, are specified in Euro code Ⅰ.

External pressures

In Euro code Ⅰ, the mean wind velocity at height z, $U(z)$, is defined by multiplying the reference wind velocity U_{bas}, representative for the climate of the site, by coefficients that take into account the effects of *terrain roughness*(地表粗糙度) and *topography*(地貌) as functions of height above ground(see Eq. 4-1):

$$U(z)=c_r(z)c_t(z)U_{bas} \tag{4-1}$$

$c_r(z)$ is the *roughness coefficient*(粗糙度系数) at height z defined by the *logarithmic*(对数的) profile(see Eq. 4-2):

$$c_r(z)=k_T\ln(z/z_0) \tag{4-2}$$

c_r is equal to one at a height of 10 m above reference terrain with a roughness length of 0.05 m. The topography coefficient, $c_r(z)$, takes into account the increase of mean wind velocity over hills and *escarpments*(悬崖,陡斜坡). For flat terrain, $c_t(z)=1$, indicating that the mean wind velocity at height z, $U(z)$, is equal to the roughness coefficient multiplied by the reference wind velocity(see Eq. 4-3):

$$U(z)=c_r(z)U_{bas} \tag{4-3}$$

The *characteristic wind pressure*(特征风压) acting on the external surfaces of a structure, F_e, is calculated as Eq. 4-4

$$F_e=q_{bas}c_e(z_e)c_{pe} \tag{4-4}$$

where q_{bas} is the *reference velocity pressure*(基准风速压) representative for the climate of the site, c_e is the *exposure coefficient*(暴露系数) which takes into account the effects of terrain roughness, topography and height above ground on the mean wind velocity and air turbulence, z_e is the reference height for the external pressure considered and C_{pe} is the external pressure coefficient specified in Euro code Ⅰ. External pressure coefficients are tabulated for areas of 1 and 10 m^2, respectively, and coefficients for intermediate surface areas are found using a *logarithmic interpolation*(对数内插法) based on area. The lack of pressure correlation over surfaces larger than 1 m^2 is thereby taken into account.

The exposure coefficient at height z, $c_e(z)$, is defined as

$$c_e(z)=\frac{q(z)[1+2k_p I_u(z)]}{q_{bas}} \tag{4-5}$$

where the ratio between the mean velocity pressure at height z, $q(z)$, and the reference velocity pressure, q_{bas}, is given by the equation below(see Eq. 4-6):

$$\frac{q(z)}{q_{bas}}=c_r^2(z)c_t^2(z) \quad 4\text{-}6$$

The exposure coefficient is seen to be the ratio between the characteristic velocity pressure $q_{max}=(1+2k_p I_u)$(see Eq. 4-4), and the reference velocity pressure q_{bas}. The external pressure coefficient, c_{pe}, could be interpreted as the ratio between the characteristic external wind pressure, F_e, and the characteristic velocity pressure q_{max}.

The characteristic velocity pressure used for static wind load, i. e. the "simple" method stated in Euro code Ⅰ, is arbitrarily based on a peak factor of 3. 5. Other peak factor choices would have given alternative specifications of external pressure coefficients and exposure coefficients, but the wind pressure F_e should not, of course, be influenced by the arbitrary peak factor chosen for codification. For more wind-sensitive structures, such as buildings over 200 m high or bridges with spans greater than 200 m, and for all structures deemed to be dynamically sensitive, as defined by the value of a dynamic coefficient, this assumption of a peak factor of 3. 5 is not adopted in Euro code Ⅰ.

The external pressure coefficients given in Euro code Ⅰ are based on results obtained in experiments, typically wind-tunnel tests. Extreme pressures and suctions measured in the experiments are normalized as shown in Eq. 4-4 in order to determine the external pressure coefficients specified. Coefficients equal to the ratio between mean pressures/mean suctions and mean velocity pressures are often different from the external pressure coefficients specified in Euro code Ⅰ. The external pressure coefficients c_{pe} give no detailed information on mean pressures and mean suctions on the structure.

For flat roofs, *monopitch roofs*(单坡屋顶), *duopitch roofs*(双坡屋顶), hipped roofs and multiplane roofs, the height of the highest roof point is chosen as the reference height, z_e. For the walls of rectangular buildings, the reference height specified depends on the building aspect ratio h/b. For low buildings where $h< b$, the roof height is chosen as the reference. For taller buildings, the wall is divided into different regions, each with a specific reference height.

In Euro code Ⅰ, the mean wind velocity at height z, $U(z)$, is defined by multiplying the reference wind velocity U_{bas}, representative for the climate of the site, by coefficients that take into account the effects of terrain roughness and topography as functions of height above ground.

Reading Skills(7)—SQ3R

A useful technique for reading a textbook assignment is called SQ3R—survey, question, read, *recite*(背诵,朗读), and review. It describes the five successive steps that should be followed while reading study-type material.

1. Survey: When you are assigned a section of a textbook to study, first survey the pages to get a general idea of the material. Skim quickly over the topic headings; look at the pictures, graphs, charts, or diagrams; see if there are questions or a summary at the end.

2. Question: After a rapid survey, ask yourself questions based on the material you have surveyed. This helps you to read with a purpose, looking for specific answers and *anticipating*(预期) essential points of information.

3. Read: Next read as rapidly as possible. Because you know what you are looking for and where you are going, your reading speed should be faster than you knew the questions before.

4. Recite: At the end of each section, summarize the material by reciting to yourself the important points. This helps you *consolidate*(巩固) the information you have read, to relate it to previous information, and to prepare yourself for what will follow.

5. Review: Finally, when you have finished the assignment, immediately review the material so that it will form a unified whole. Also, when you have the next assignment in the text, review the preceding material, surveying it rapidly to refresh your memory. Each section, though read separately and at different times, will fit together into the total organization of the material that the author intended.

Try to use the SQ3R technique by reading the above reading material.

Lesson 5

SERVICEABILITY

The member has been assumed to be at a hypothetical overload state for safety purpose. It is also important that member performance in normal service be satisfactory, when loads are those actually expected to act, i. e. , when load factors are 1. 0. This is not guaranteed simply by providing adequate strength. Serve load deflections under full load may be excessively large, or long-term deflections due to sustained loads may cause damage. Tension cracks in beams may be wide enough to be visually disturbing, and in some cases may reduce the durability of the structure. These and other questions, such as vibration or fatigue, require consideration.

Serviceability studies are carried out based on elastic theory, with stresses in both concrete and steel assumed to be proportional to strain. The concrete on the tension side of the neutral axis may be assumed uncracked, partially cracked, or fully cracked, depending on the loads and material strengths.

In early reinforced concrete designs, questions of serviceability were dealt with indirectly, by limiting the stresses in concrete and steel at service loads to the rather conservative values that had resulted in satisfactory performance. In contrast, with current design methods that permit more slender members through more accurate assessment of capacity, and with higher-strength materials further contributing to the trend toward smaller member sizes, such indirect methods no longer work. ① The current approach is to investigate service load cracking and deflections specifically, after proportioning members based on strength requirements.

In this chapter, members will be developed to ensure that the cracks associated with flexure of reinforced concrete beams are narrow and well distributed, and that short and long-term deflections at loads up to the full service load are not objectionably large.

Cracking of flexural members

All reinforced concrete beams crack, generally starting at loads well below service level, and possibly even prior to loading due to restrained shrinkage. Flexural cracking due to loads is not only inevitable, but actually necessary for the reinforcement to be used effectively. Prior to the formation of flexural cracks, the steel stress is no more than n times the stress in the adjacent concrete, where n is the modular ratio E_s/E_c. For materials common in current practice, n is approximately

8. Thus when the concrete is close to its modulus of rupture of about 500 psi, the steel stress will be only 8×500 psi$=4\ 000$ psi, far too low to be very effective as reinforcement. At normal service loads, steel stresses of 8 or 9 times that value can be expected.

In a well-designed beam, flexural cracks are fine, so-called hairline cracks, almost invisible to the casual observer, and they permit little if any corrosion of the reinforcement. As loads are gradually increased above the cracking load, both the number and width of cracks increase, and at service load level a maximum width of crack of about 0.016 in. is typical. If loads are further increased, crack widths increase further, although the number of cracks is more or less stable.

Cracking of concrete is a random process, highly variable and influenced by many factors. Because of the complexity of the problem, present methods for predicting crack widths are based primarily on test observations. Most equations that have been developed predict the probable maximum crack width, which usually means that about 90 percent of the crack widths in the member are below the calculated value.② However, isolated cracks exceeding twice the computed width can sometimes occur.

Variables affecting width of cracks

In the discussion of the importance of a good bond between steel and concrete, it was pointed out that if proper end anchorage is provided, a beam will not fail prematurely, even though the bond is destroyed along the entire span. However, crack widths will be greater than for an otherwise identical beam in which good resistance to slip is provided along the length of the span. In general, beams with smooth round bars will display a relatively small number of rather wide cracks in service while beams with good slip resistance ensured by proper surface deformations on the bars will show a larger number of very fine, almost invisible cracks. Because of this improvement, reinforcing bars in current practice are always provided with surface deformations, the maximum spacing and minimum height of which are established by ASTM specifications A 615, A 706, and A 996.

A second variable of importance is the stress in the reinforcement. Studies by Gergely and Lutz and others have confirmed that crack width is proportional to f_s^n, where f_s is the steel stress and n is an exponent that varies in the range from about 1.0 to 1.4. For steel stresses in the range of practical interest, say from 20 to 36 ksi, n may be taken equal to 1.0. The steel stress is easily computed based on elastic cracked-section analysis. Alternatively, f_s may be taken equal to $0.60f_y$ according to ACI code 10.6.4.

Experiments by Broms and others have shown that both crack spacing and crack width are related to the concrete cover distance d_c, measured from the center of the bar to the face of the concrete. In general, increasing the cover increases the spacing of cracks and also increases crack width. ③ Furthermore, the distribution of the reinforcement in the tension zone of the beam is important. Generally, to control cracking, it is better to use a larger number of smaller-diameter bars to provide the required A_s than to use the minimum number of larger bars, and the bars should be well distributed over the tensile zone of the concrete. For deep flexural members, this includes additional reinforcement on the sides of the web to prevent excessive surface crack widths above the level of the main flexural reinforcement.

Words and Expressions

hypothetical	[ˌhɑipəˈθetikəl]	*adj*. 假设的,假定的,爱猜想的
deflection	[diˈflekʃən]	*n*. 挠度,偏差
crack	[kræk]	*n*. 裂缝,缺点
		v. 裂开,破裂,变声
durability	[ˌdjuərəˈbiləti]	*n*. 持久性(耐久性)
fatigue	[fəˈtiːg]	*n*. 疲乏,疲劳
		v. 使……疲劳
assessment	[əˈsesmənt]	*n*. 估价,评估
flexure	[ˈflekʃə]	*n*. 弯曲,褶缝,屈曲
flexural	[ˈflekʃərəl]	*adj*. 弯曲的(挠曲的)
restrained	[riˈstreind]	*adj*. 克制的,自制的,受约束的
shrinkage	[ˈʃrinkidʒ]	*n*. 收缩,缩小,减少
inevitable	[inˈevitəbl]	*adj*. 不可避免的,必然(发生)的
rupture	[ˈrʌptʃə(r)]	*n*. 破裂,断裂
		v. 使破裂
isolate	[ˈaisəleit]	*adj*. 孤立的
		v. 使隔离,使孤立
bond	[bɔnd]	*n*. 黏结剂,黏合剂
anchorage	[ˈæŋkəridʒ]	*n*. 锚具,锚固,停泊地点,抛锚地点
prematurely	[ˌpriːməˈtjuəli]	*adv*. 过早地,早熟地
slip	[slip]	*n*. 滑移
		v. 滑移
spacing	[ˈspeisiŋ]	*n*. 间距
exponent	[eksˈpəunənt]	*n*. 指数,说明者,说明物
		adj. 说明的
web	[web]	*n*. 网,腹板
long-term deflection		长期挠度
sustained load		长期载荷
neutral axis		中和轴,中性轴

service load	使用荷载
hairline crack	毛细裂缝，发裂，发纹
crack width	裂缝宽度
round bar	圆棒，圆钢，圆钢筋
deep flexural members	深受弯构件

Notes

① In contrast, with current design methods that permit more slender members through more accurate assessment of capacity, and with higher-strength materials further contributing to the trend toward smaller member sizes, such indirect methods no longer work.

本句中 In contrast 作为状语，意思是"作为对照"。with current design methods 和后面的 and with higher-strength materials 作为状语，是谓语 no longer work 的原因。在句子中，that 引导的句子是其前面 design methods 的定语从句，contributing to 是现在分词短语，作为 matcrials 的定语。

② Most equations that have been developed predict the probable maximum crack width, which usually means that about 90 percent of the crack widths in the member are below the calculated value.

本句中，Most equations 为主语，that have been developed… 为其定语从句；predict the crack width 为谓语和宾语，which usually means 为宾语的定语从句，其后的 that about… 是 means 的定语从句。

③ In general, increasing the cover increases the spacing of cracks and also increases crack width.

句中 increasing the cover 为主语，increases the spacing of cracks and also increases crack width 中两个 increases 为谓语。

Exercises

Ⅰ. Translate the following words into Chinese.

1. member performance ____________________
2. rational safety provisions ____________________
3. concrete cover distance ____________________
4. random variable ____________________
5. deep flexural members ____________________

Ⅱ. Translate the following words into English.

1. 结构的适用性____________________
2. 结构的耐久性____________________
3. 裂缝宽度____________________
4. 长期挠度____________________
5. 使用荷载____________________

Ⅲ. Translate the following sentences into Chinese.

1. Tension cracks in beams may be wide enough to be visually disturbing, and in some cases may reduce the durability of the structure.
2. All reinforced concrete beams crack, generally starting at loads well below service level, and possibly even prior to loading due to restrained shrinkage.
3. In early reinforced concrete designs, questions of serviceability were dealt with indirectly, by limiting the stresses in concrete and steel at service loads to the rather conservative values that had resulted in satisfactory performance.
4. In the discussion of the importance of a good bond between steel and concrete, it was pointed out that if proper end anchorage is provided, a beam will not fail prematurely, even though the bond is destroyed along the entire span.
5. Experiments by Broms and others have shown that both crack spacing and crack width are related to the concrete cover distance d_c, measured from the center of the bar to the face of the concrete.

Ⅳ. Translate the following sentences into English.

1. 在设计良好的钢筋混凝土梁中,裂缝须足够小,以防出现超过规范允许宽度的钢筋锈蚀。
2. 为控制裂缝,最好使用直径小而数量多的钢筋,而不是大直径而数量少的钢筋。
3. 一般地,配置光圆钢筋的梁,裂缝会表现得少而宽;而配置适量变形钢筋的梁,裂缝数量多而宽度细微到几乎不可见。
4. 对深受弯构件,除受弯主筋外,需在其腹部两侧配置附加钢筋以防其表面裂缝宽度超限。

Ⅴ. Answer the following questions briefly.

1. What affects width of cracks? Is this true "failure will never occur in a well designed structure"?
2. Is that true "all reinforced concrete beams crack"?
3. What does tension crack affect if its width is wide enough?

Reading Materials

Control of Deflections

In addition to limitations on cracking, it is usually necessary to impose certain controls on deflections of beams to ensure serviceability. Excessive deflections can lead to cracking of supported walls and partitions, ill-fitting door and windows, poor roof drainage, misalignment of sensitive machinery and equipment, or visually offensive *sag*(下弯). It is important, therefore, to maintain control of deflections, in one way or another, so that members designed mainly for strength at prescribed overloads will also perform well in normal service.

There are presently two approaches to deflection control. The first is indirect and consists in setting suitable upper limits on the *span-depth ratio*(跨高比). This is simple, and it is satisfactory in many cases where spans, loads and load distributions, and member sizes and proportions fall in the usual ranges. Otherwise, it is essential to calculate deflections and to compare those predicted values with specific limitations that may be imposed by codes or by special requirements.

It will become clear, in the sections that follow, that calculations can, at best, provide a guide to probable actual deflections. This is so because of uncertainties regarding material properties, effects of cracking, and load history for the member under consideration. Extreme precision in the calculations, therefore, is never justified, because highly accurate results are unlikely. However, it is generally sufficient to know, for example, that the deflection under load will be about 1/2 in. rather than 2 in. , while it is relatively unimportant to know whether it will actually be 5/8 in. rather than 1/2 in.

The deflections of concern are generally those that occur during the normal service life of the member. In service, a member sustains the full dead load, plus some fraction or all of the specified service line loads. Safety provisions of the ACI code and similar design specifications ensure that, under loads up to the full service load, stress in both steel and concrete remain within the elastic ranges. Consequently, deflections that occur at once upon application of load, the so-called immediate deflections, can be calculated based on the properties either of the uncracked elastic member, the cracked elastic member, or some combination of these.

It was pointed out before that in addition to concrete deformations that occur immediately when load is applied, there are other deformations that take place gradually over an extended period of time. These time-dependent deformations are chiefly due to concrete creep and shrinkage. As a result of these influences, reinforced concrete members continue to deflect with the passage of time. Long-term deflections continue over a period of several years, and may eventually be two or more times the initial elastic deflections. Clearly, methods for predicting both instantaneous and time-dependent deflections are essential.

Immediate deflections

Elastic deflections can be expressed in the general form,

$$\Delta=\frac{f(\text{loads},\text{spans},\text{supports})}{EI} \tag{5-1}$$

Where EI is the flexural rigidity and f(loads, spans, supports) is a function of the particular load, span, and support arrangement. For instance, the deflection of a uniformly loaded simple beam is $5wl^4/384EI$, so that $f=wl^4/384$. Similar deflection equations have been tabulated or can easily be computed for many other loadings and span arrangements, simple, fixed, or continuous, and the corresponding f functions can be determined. The particular problem in reinforced concrete structures is therefore the determination of the appropriate flexural rigidity EI for a member consisting of two materials with properties and behavior as widely different as steel and concrete.

If the maximum moment in a flexural member is so small that the tensile stress in the concrete does not exceed the modulus of rupture f_r, no flexural tension cracks will occur. The full, uncracked section is then available for resisting stress and providing rigidity. In agreement with this analysis, the effective moment of inertia for this low range of loads is that of the uncracked transformed section I_{ut}, and E is the modulus of concrete E_c. Correspondingly, for this load range,

$$\Delta_{iu}=\frac{f}{E_c I_{ut}} \qquad 5\text{-}2$$

At higher loads, flexural tension cracks are formed. In addition, if shear stresses exceed v_{cr} and web reinforcement is employed to resist them, diagonal cracks can exist at service loads. In the region of flexural cracks, the position of the neutral axis varies: directly at each crack it is located at the level calculated for the cracked transformed section; midway between cracks it dips to a location closer to that calculated for the uncracked transformed section. Correspondingly, flexural-tension cracking causes the effective moment of inertia to be that of the cracked transformed section in the immediate neighborhood of flexural-tension cracks, and closer to that of the uncracked transformed section midway between cracks, with a gradual transition between these extremes.

Deflections due to *long-term loads*(长期荷载)

Initial deflections are increased significantly if loads are sustained over a long period of time, due to the effects of shrinkage and creep. These two effects are usually combined in deflection calculations. Creep generally dominates, but for some types of members, shrinkage deflections are large and should be considered separately.

It was pointed out that creep deformations of concrete are directly proportional to the compressive stress up to and beyond the usual service load range. They increase *asymptotically*(渐近地) with time and, for the same stress, are larger for low-strength than for high-strength concretes. The ratio of additional time-dependent strain to initial elastic strain is given by the creep coefficient C_{cu}.

Reading Skills(8)—Context Clues

Efficient reading requires the use of various problem-solving skills. For example, it is impossible for you to know the exact meaning of every word you read, but by developing your guessing ability, you will be able to understand enough to arrive the total meaning of a sentence, paragraph, or essay. These exercises are designed to help you improve your ability to guess the meaning of unfamiliar words by using context clues. (Context refers to the sentence and paragraph in which a word occurs.) In using the context to decide the meaning of a word you have to use your knowledge of grammar and your understanding of the author's ideas. Although there is no formula which you can memorize to improve your ability to guess the meaning of unfamiliar words, you should keep the following points in mind.

①Use the meanings of the other words in the sentence (or paragraph) and the meaning of the sentence as a whole to reduce the number of possible meanings.

②Use grammar and punctuation clues which point to the relationships among the various parts of the sentence.

③Be content with a general idea about the unfamiliar words the exact definition or synonym is not always necessary.

④Learn to recognize situations in which it is not necessary to know the meaning of the word.

Read the following materials carefully. Concentrate on developing your ability to guess the meaning of unfamiliar words(italic ones) using context clues.

Mathematical Modeling of the Structural Components and Discretization of Bridge Structure

In the mathematical modeling of structural *components*, it is *current* practice to obtain *discretized analytical* elements that *deform* the same way when subjected to loading as the real structural component. This implies that the model element must have the same *stiffness properties* as the structure member.

When it comes to the modeling of the whole structure, the *manner* of discretization of the structural members into an adequate number of elements has to be decided. The following factors influence this decision.

①Type of structural *configuration*, i. e. arch structure, continuous box-girder on *monolithic* supports, etc. and the nature of the interaction between members.

②Type of loading to which the structure will be subjected, i. e. dead and live loads acting in only one plane, or earthquake loading in any direction.

③The *nature* of output required, i. e. longitudinal beam action only or beam action and *transverse* slab bending or stress concentrations, etc.

④Number of output points required. Most programs supply stress *resultants* only at the end of members, so that finer *subdivision* may be necessary for particular load types.

⑤In finite element analysis, it must be determined how many elements are required for a sufficient solution. Unless the particular elements types have been used before, it is essential to test a *similar* mesh *configuration* with known results before using them in the bridge model.

Lesson 6

GENERAL CHARACTERISTICS OF EARTHQUAKE-RESISTANT STRUCTURES

Structure types

Structures that are continuous (see Fig. 6-1(a)) in nature and more or less uniformly distributed throughout a building generally perform well when subjected to earthquakes. The primary reason for this is that the earthquake resistance of a structure is dependent, to a large extent, on its ability to absorb the energy input associated with ground motions. Pin-connected structures, such as traditional post-and-beam assemblies (see Fig. 6-1(b)), are far less capable of absorbing energy than are comparable continuous structures (e. g. frames with monolithic joints). ①

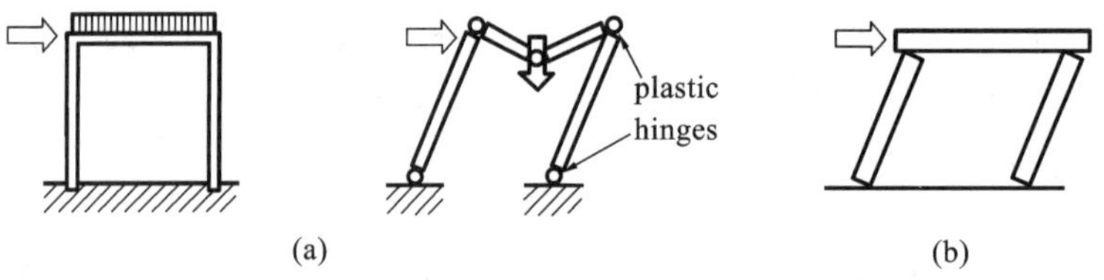

Fig. 6-1 Continuous structures and pin-connected ones

(a) frame; (b) post-and-beam assembly

The formation of plastic hinges in framed structures (which must precede their collapse) requires a significant energy input. Continuous structures, therefore, are quite effectively used in buildings in earthquake-hazard zones. Either steel or cast-in-place reinforced concrete framed structures designed to be ductile under earthquake forces can be used. Precast concrete structures, with their lack of continuity at connections, are more difficult, although not impossible, to make work.

Even within the category of continuous structures, however, differences exist. Fig. 6-2 illustrates the plans of two structures deriving their ability to carry lateral forces from frame action. The first structure, illustrated in Fig. 6-2(a), with frames distributed throughout the structure, is generally preferable to that shown in Fig. 6-2(b), which has frames around the periphery only. ② The first structure has greater redundancy, and consequently has greater reserve strength than the second structure. The failure of relatively few members in the outer plane of the structure shown in Fig. 6-2(b) may lead to total collapse, whereas many more members must fail in the more redundant structure for it to collapse. Generally, structures with redundancy are preferable to those without redundancy.

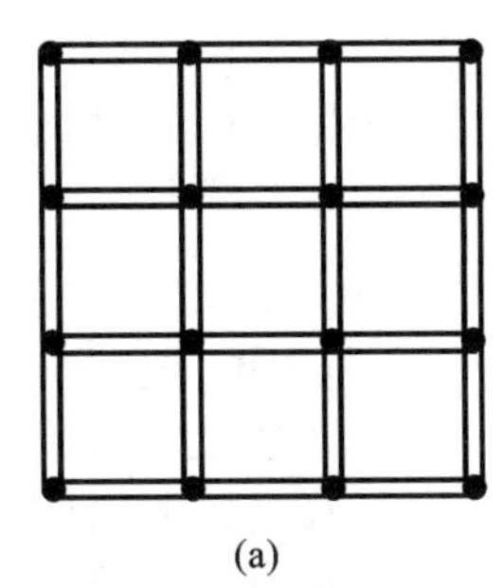

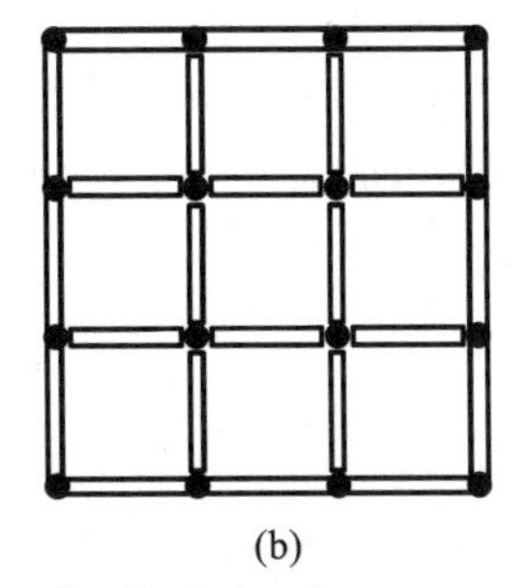

(a) (b)

Fig. 6-2 Redundant versus nonredundant structures

(a) highly redundant structure: frames are used throughout the building;

(b) structure with more limited redundancy: frames are on the periphery of the building only

Another general characteristic of viable earthquake-resistant structures is that column-and-beam elements are generally coaxial. Offsets or nonaligned members often present extremely difficult design problems.

Earthquake-resistant structures also typically have floor and roof planes designed as rigid diaphragms capable of transmitting internal forces to lateral-load-resisting elements through beam-like action (see Fig. 6-3). Fig. 6-3(a) shows that the horizontal plane acts like a beam in carrying earthquake-induced forces to shear walls or other lateral-load-carrying mechanisms; if diaphragms are improperly designed, failure can result in floor or roof planes(see Fig. 6-3(b)).

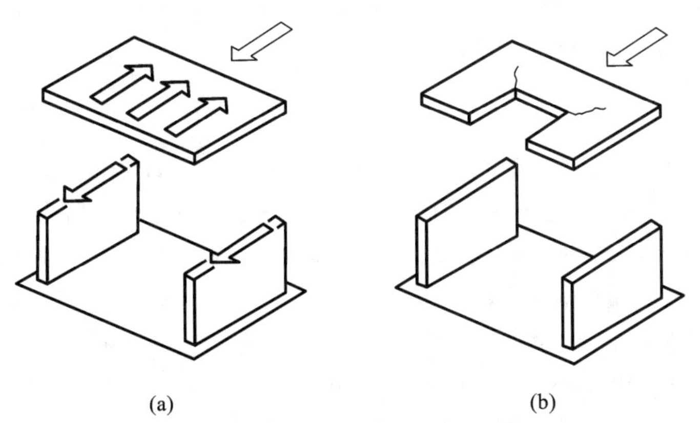

(a) (b)

Fig. 6-3 Importance of rigid floor and roof elements

(a) typical diaphragm action; (b) failure in floor or roof planes

Another aspect of earthquake-resistant structures is that they are designed such that horizontal elements which fail due to earthquake motions do so before any vertical members fail (see Fig. 6-4). ③ Failure should never occur in vertical members first. The reason for this is simply consideration of life-saving. Horizontal elements in continuous structures (e. g. slabs or beams) rarely fall down completely, even after receiving extreme damage, and when they do, the collapse is fairly localized. (This is, in general, not true of pin-connected horizontal members.) When columns receive damage, complete collapse is imminent. The collapse of a column generally causes other portions of a structure to collapse as well. The effects of a single column collapsing can be extensive.

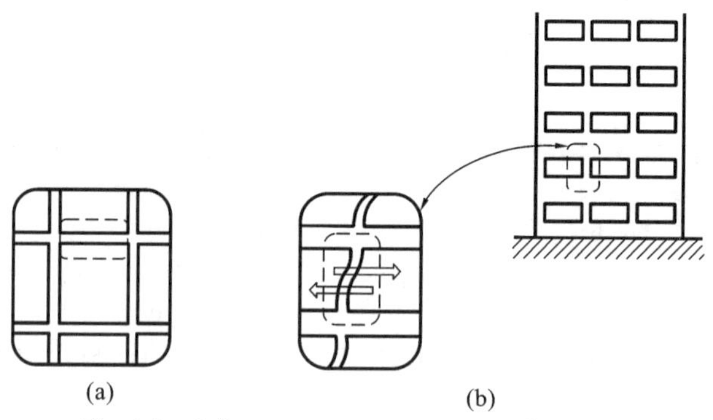

Fig. 6-4 A "strong-column, weak-beam" strategy

(a) beam failure occurs first; (b) column failure occurs first (very undesirable)

To ensure that horizontal elements fail first, care is usually exercised in the design and general proportioning of beam-and-column elements. Extremely deep beams (e. g. spandrels) on light columns are generally best avoided, since experience indicates that such buildings often receive significant damage in the light columns, which have to pick up all the laterally acting forces by shear and bending. In contrast, shear walls with a series of spaced, small openings tend to perform somewhat better. Good engineering design, it should be noted, can make all types work adequately.

Materials

As noted before, fully continuous structures are desirable for use in earthquake-prone regions. Coupled with the idea of continuity is that of ductility. For energy absorption to take place, ductility is essential. Steel is a naturally ductile material and is consequently often used. Cast-in-place reinforced concrete can also be made to have a high degree of ductility by carefully controlling member proportions and the amount and placement of reinforcing steel. Reinforced-concrete buildings of this type are frequently used in earthquake zones. Precast concrete structures, however, can be difficult to design for safety in earthquake zones because of the problems involved in achieving a continuous, ductile structure. Such structures are also typically high-mass structures. The high mass contributes to the magnitude of the seismic forces developed and compounds the problem of their use.

Timber can be an extremely good material for use in earthquake regions. It is lightweight and capable of absorbing large amounts of energy when deformed and before collapse. Low-rise wood-framed structures are highly earthquake resistant and perform well in earthquake regions.

Words and Expressions

frame	[freim]	*n.* 框，结构，骨架 *v.* 构筑，设计
monolithic	[ˌmɔnəˈliθik]	*n.* 整体式的，独块巨石的，整体浇筑的
joint	[dʒɔint]	*n.* 接缝，节点
ductile	[ˈdʌktail]	*adj.* 有延性的，易变形的
precast	[priːˈkaːst]	*adj.* 预制的
assembly	[əˈsembli]	*n.* 集合，集会，装配
lateral	[ˈlætərəl]	*adj.* 侧向的，横向的
preferable	[ˈprefərəbl]	*adj.* 更可取的，更好的，更合意的
coaxial	[ˌkəuˈæksəl]	*adj.* 同轴的，共轴的
imminent	[ˈiminənt]	*adj.* 即将来临的，逼近的
spandrel	[ˈspʌndrəl]	*n.* 拱肩，拱上空间，上下层窗空间
earthquake-resistant structure		抗震结构
be subjected to		使经受，使遭受
pin-connected		铰接的
plastic hinge		塑性铰
earthquake-hazard zone		地震受灾区
cast-in-place		现场浇筑的
preferable to		比……可取，优于，胜过
rigid diaphragms		刚性隔板，刚性楼板
strong-column		强柱
weak-beam		弱梁
earthquake-prone region		易发生地震的地区
high-mass structure		重量大的结构

Notes

① Pin-connected structures, such as traditional post-and-beam assemblies (see Fig. 6-1(b)), are far less capable of absorbing energy than are comparable continuous structures (e. g. frames with monolithic joints).
本句中 such as traditional post-and-beam assemblies 为插入语，整句话可以译为"铰接连接的结构，例如传统的立柱和梁的组合体，与连续结构相比，其吸收能量的能力小多了"。

② The first structure, illustrated in Fig. 6-2(a), with frames distributed throughout the structure, is generally preferable to that shown in Fig. 6-2(b), which has frames around the periphery only.
句中 illustrated in Fig. 6-2(a), with frames distributed throughout the structure 为插入语，修饰其前的 structure；is generally preferable to that 中 that 指代 structure；shown in Fig. 6-2(b)为过去分词短语作 that 的定语。

③ Another aspect of earthquake-resistant structures is that they are designed such that horizontal elements which fail due to earthquake motions do so before any vertical members fail (see Fig. 6-4).

…that they are …为表语从句，在这个表语从句中…such that…为目的状语，that 后面引导的为状语从句。

Exercises

Ⅰ. Translate the following words into Chinese.

1. earthquake-resistant structures ____________
2. post-and-beam assemblies ____________
3. lateral-load-resisting elements ____________
4. rigid diaphragms ____________
5. structures with redundancy ____________
6. timber structures ____________

Ⅱ. Translate the following words into English.

1. 预制混凝土结构____________
2. 能量的吸收____________
3. 地震受灾区____________
4. 现浇结构____________
5. 水平抗侧力构件____________
6. 塑性铰____________
7. 延性结构____________

Ⅲ. Translate the following sentences into Chinese.

1. The primary reason for this is that the earthquake resistance of a structure is dependent, to a large extent, on its ability to absorb the energy input associated with ground motions.
2. The failure of relatively few members in the outer plane of the structure shown in Fig. 6-2(b) may lead to total collapse, whereas many more members must fail in the more redundant structure for it to collapse.
3. Earthquake-resistant structures also typically have floor and roof planes designed as rigid diaphragms capable of transmitting internal forces to lateral-load-resisting elements through beam-like action.
4. Horizontal elements in continuous structures (e. g. slabs or beams) rarely fall down completely, even after receiving extreme damage, and when they do, the collapse is fairly localized.
5. Precast concrete structures, however, can be difficult to design for safety in earthquake zones because of the problems involved in achieving a continuous, ductile structure.

Ⅳ. **Translate the following sentences into English.**

1. 在经受地震的时候，连续结构往往表现良好。
2. 结构抵抗地震的能力很大程度上取决于它吸收输入能量的能力。
3. 预制混凝土结构，由于其节点缺乏连续性，不适用于地震区。
4. 作为其特色，抗震结构往往拥有楼板或屋面板作为刚性隔板，以将内力传递到水平抗侧力构件上。

Ⅴ. **Answer the following questions briefly.**

1. What kinds of structure are suitable for earthquake zones? Why?
2. What kinds of structure material are suitable for earthquake zones? Why?
3. Do all continuous structures behave the same in energy absorbing?
4. Why does plastic hinge play so important a role in structures?
5. What's the function of floors in earthquake-zone structures?

Reading Materials

The Structural Engineering of the Timber Grid Shell

The location and reason for building

At the Weald & Downland Open Air Museum, in Sussex, England, more than 45 historic buildings from South East England have been rescued and rebuilt. The Museum wanted a new building that would stand as a *testament*(实证，遗嘱) to architectural and building techniques of the early 21st century. The resulting structure incorporates a wide range of *carpentry*(木工) disciplines and skills, but, most significantly, the Downland *grid shell*(网壳) is the first structure to be built in the UK using techniques similar to those used for the Mannheim grid shell in 1975.

The building commissioned by the Museum is a two-storey structure. The upper storey is a workshop. It is the workshop roof that is a timber grid shell. In principle, the construction sequence of the gird shell is simple: a regular grid of slender timber *laths*(木板条) is laid out flat and is subsequently shaped into a doubly *curved*(曲线的) form. It is possible to do this because of the flexibility of the laths and the rotational freedom at the grid shell *intersections*(交叉点). This is in contrast to a steel or concrete grid shell that is erected in its doubly curved shape, with the difficulty of *prefabricating*(预制) hundreds or thousands of individual, different nodes. After forming the shell, shear stiffness is provided by adding cross *bracing*(支撑) to the doubly curved grid of timber laths.

Precedents for the building

The first double-layered timber grid shell was constructed for the Mannheim Bundesgartenschau in 1975. The Mannheim grid shell was constructed by laying out the *flat*(平坦的，扁平的) *lattice*(网架) on the ground and then pushing it up using *spreader beams*(分配梁) and *scaffolding*(脚手架) towers at strategic locations. In transforming the flat lattice to the final shape there was a significant component of lateral movement related to the gain in height. To accommodate this the scaffolding towers were moved using *forklifts*(铲车,叉式升降机). The building has been very successful and remains as one of the finest buildings of the 20th century, despite its original two-year design life. The process of pushing up on discrete areas during erection did concentrate stress within particular areas, the consequence of which was a notable number of *breakages*(破损量) of laths and *finger joints*(指形接合). It also required *operatives*(有经验的工人) to work at height and under the temporarily supported grid; a scenario that would not be permitted under today's CDM regulations.

More recently Buro Happold have designed two single-layer grid shells, one in timber, the other in *cardboard*(纸板). The first were the grid shell sculptures at the Earth Centre in Doncaster constructed in 1998(see Fig. 6-5). These were small timber lattice structures composed of a single layer of thin timber laths. A *crane*(起重机,吊车) was used to lift the lattice into position and then it was manipulated with *struts*(压杆) and *ties*(拉杆) from a flat grid to the final shape. An important lesson learnt on this project was that the nodes need to be extremely loose to enable rotation during formation.

Fig. 6-5 The grid shell at the Earth Centre in Doncaster

The second recent project was the Japanese Pavilion at the Hanover Expo 2000. This building was similar in shape to the Downland grid shell and composed of cardboard tubes(see Fig. 6-6). The erection used the *modular*(模数的,制成标准尺寸的) scaffold system PERI-UP, shown in Fig. 6-7. The flat mat was laid out on a low-level scaffold bed and was pushed up into position using proprietary PERI *jack*(千斤顶) known as a MULITPROP. Due to the similarity of the shape of this shell with the Downland grid shell, a great deal was learnt from the experience of erecting the Hanover building.

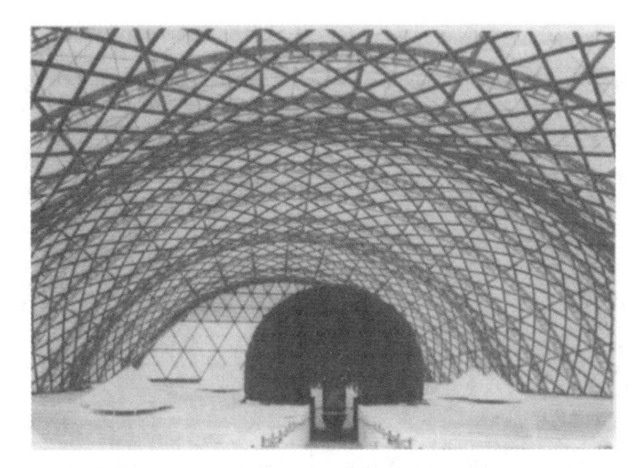

Fig. 6-6 The Japanese Pavilion, Hanover Expo 2000

Fig. 6-7 Erection of Hanover grid shell using PERI scaffolding

Constructions

The benefit of timber grid shells becomes apparent in the construction stage. Scaffolding supplied by the German company PERI was used in the construction of the Japanese Pavilion where it proved to be very suitable and effective. For the Downland grid shell, PERI were able to bring their experience to bear in deciding the best procedure for formation.

Layout of flat mat

The lattice was laid out as a flat mat (see Fig. 6-8) composed of squares with a 1 m edge length; the resulting lattice mat, which would be 47 m long×25 m wide if laid out *orthogonally*(互相垂直的), was stretched longitudinally to achieve 50 m length of the completed structure.

An innovation was the choice of layout level. Instead of laying out the mat at or near the floor level, as had been done on previous grid shells, a birdcage of scaffolding with a plan area the size of the layout mat was built to a height of 7 m above the workshop floor.

Fig. 6-8 The completed layout of the flat lattice mat

Reading Skills(9)—Stems and Affixes(Ⅰ)

In the chart below, some of the most common prefixes are listed *alphabetically*(按字母顺序地). The meaning of each prefix is given as an area of meaning because most often there is no one single specific meaning. In the right-hand column, space has been left for you to record examples of words which use the prefix.

Your dictionary will provide examples. Choose those which are familiar or potentially useful to you and write down them in the column.

Prefix	Area of meaning	Examples
ab-	away, from, down	
anti-	against, opposite	
auto-	self	
bi-	two, twice	
circum-	around	
con-, co-	with, together	
col-, com-		
de-	down, reversing	
dis-	not, opposite of	
ex-, e-	out, from	
im-	in, not	
in-	in, not	
inter-	between, among	
macro-	large	
micro-	small	
mis-	wrong, unfavorable	
mono-	one, alone	

续表

Prefix	Area of meaning	Examples
ob-	against, toward	
per-	through, thoroughly	
post-	behind, after	
pre-, prim-	first	
pro-	for, before, forward	
re-	again, less, than	
sub-	under, less, than	
trans-	across	
tri-	three	
uni-	one	
un-, ir-	not	
ultra-	beyond, excessive	

Lesson 7

CIVIL ENGINEERING MATERIALS

Civil engineering materials can be natural and manmade. They contain rock, metal, timber, concrete, soil, polymer, brick and block etc. Besides these traditional materials, new types of constructional materials are also investigated and developed and will be applied gradually.

The modern civil engineer needs to deal with a variety of materials that are often integrated in the same structure, such as steel and concrete, or are used separately for construction projects, such as pavements from asphalt and Portland cement concrete. Many of these construction materials have been with us for centuries, like timber; while others, like Portland cement and steel, are relatively new and have been used mainly during the last century. Nowadays green civil engineering materials and even eco-materials for civil engineering are recommended based on the consideration of sustainable development. The following are some introduction of principal civil engineering materials.

Metals

Metals consist of ferrous metals and nonferrous metals. The applications of metals in civil engineering are wide and varied, ranging from their use as main structural materials to their use for fastenings and bearing materials. As main structural materials, cast iron and wrought iron have been superseded by rolled-steel sections. Steel is also of major importance for its use in reinforced and prestressed concrete.

The properties of metals which make them unique among constructional materials are high tensile strength, the ability to be formed into plate, sections and wire, and the weldability. When in service, metals frequently have to resist not only high tensile or compressive stress and corrosion, but also conditions of shock loading and low temperatures. The importance of metals as constructional materials is almost invariably related to their load bearing capacity in either tension or compression and their ability to withstand limited deformation without fracture. ①

Many ferrous and nonferrous metals and alloys are available to engineers. Iron alloys are used in large quantities than the alloys of any other metal. This arises from the relative cheapness with which steels and cast irons can be produced with a variety of useful properties.

Structural steels

Structural steels are main materials in steel structures. Design of steel structures is based primarily on the strength of the steel but ductility, toughness and weldability are often important properties. The weldability of steels deteriorates with the increase of carbon content. The weldability of steels is very important because welded structures give a weight saving and ease of fabrication compared with bolted or riveted structures.

Plain carbon steel, also called mild steel, has good ductility and weldability if its carbon content is low and high yield strength if its carbon content is high. Consequently the carbon content is limited to 0.38% maximum in the basic structural steels to give a compromise between the opposing requirements. The carbon steels also contain the elements of manganese, silicon, sulphur and phosphorus that arise and are controlled in the steelmaking process. Manganese improves strength and ductility and between 0.5% and 1.0% is normally present. Silicon improves the strength but if present in excessive amounts may reduce the strength, and so the silicon content rarely exceeds 0.6%. Sulphur and phosphorus embrittle the steel and are controlled to 0.05% maximum.

Low alloy steels contain more alloy content than plain carbon steels. The alloy elements are manganese (0.8%-1.7%), silicon (≤0.50%), vanadium (0.02%-0.20%), niobium (0.015%-0.06%), etc. Compared with the plain steels, low alloy steels have higher strength, better ductility and same weldability. Low alloy steels are usually used in tall buildings, long span buildings, high-rise structures, long span bridges, heavy industrial factories, etc.

As we know, concrete has low tensile and bending strength compared to its high compressive strength, and concrete is easy to crack even under a very low stress. Reinforced concrete can overcome the deficiencies in the tensile and bending strengths of concrete. The reinforcing steel must have adequate tensile properties and form a strong bond with the concrete. The bond force is purely mechanical and arises from surface roughness and friction of the reinforcement. ②

Mild steel with a maximum carbon content of 0.25% is suitable and supplied in three conditions. These are hot rolled, cold rolled and hard drawn.

Pre-stressing steels must have a high yield stress in tension so that a high tensile force can be inducted in them. High strength steel is mainly used for pre-stressing steels.

Aluminum and aluminum alloys

The useful engineering properties of both unalloyed and alloyed aluminum are low specific gravity, resistance to corrosion, high electrical conductivity and excellent forming properties. The low strength of aluminum is a disadvantage and for satisfactory service it must be alloyed and mechanically or thermally treated to give improved strength.

Super-purity 99.99% aluminum is too costly for general engineering applications and commercial grades, which vary in purity from 99.0% to 99.8% aluminum, is normally selected on the requirements of corrosion resistance, formability or tensile strength. Corrosion resistance and formability are enhanced and tensile strength is lowed by increased purity.

Alloy additions are chosen so that the strength can be enhanced without an adverse effect on the specific gravity or corrosion resistance. The alloying elements commonly added to aluminum are copper, manganese, magnesium, silicon, nickel and iron. Part of the addition forms a solid solution with the aluminum and part forms a compound with the aluminum or with one of the other alloying elements[③]. The corrosion resistance of an alloy is good if the solid solution and associated compound have similar chemical properties.

Concrete

Concrete is a man-made composite, the major constituent of which is natural aggregate (such as gravel and sand) and binding medium (such as cement paste, bitumen and polymers). The binding medium is used to bind the aggregate particles together to form a hard composite material.

Normal concrete has a comparatively low tensile strength and for structural applications it is normal practice either to incorporate steel bars to resist any tensile forces (steel reinforced concrete) or to apply compressive forces to the concrete to counteract these tensile forces (pre-stressed concrete or post-stressed concrete). Concrete is used structurally in buildings, shell structures, bridges, sewage treatment works, railway sleepers, roads, cooling towers, dams, chimneys, harbors, off-shore structures, coastal protection works and so on. It is used also for a wide range of pre-cast concrete products which include concrete blocks, cladding panels, pipes and lamp-standards.

The impact strength, as well as the tensile strength, of normal concrete is low and this can be improved by the introduction of randomly orientated fibers into the concrete. [④] Steel, polypropylene, asbestos glass and carbon fibers have all been used with some success in pre-cast products and in-situ concretes, including pipes, building panels and piles.

Concrete requires little maintenance and has good fire resistance. Concrete has other properties which may on occasions be considered less desirable, for example, the time-dependent deformations associated with drying shrinkage and other related phenomena.

Masonry

The term masonry is used to describe a variety of formations consisting of separate elements bonded together by some binding filler. The elements may be cut or rough stone, fired clay tile or bricks, or cast units of concrete. The binder is traditionally cement-lime mortar, although considerable effort is being made in experimentation with various new adhesive compounds. The resulting assemblage is similar to a concrete structure and possesses many of the same properties. A major difference is that the construction process does not usually require the same amount of temporary forming and bracing as it does for a structure of poured concrete. However, it requires considerable hand labor, which imposes some time limitations and makes the end-product highly subject to the individual skill of the craftsperson. Reinforcing techniques have been developed in recent years to extend the structural possibilities of masonry, especially improved its resistance to earthquakes.

Shrinkage of the mortar and thermal-expansion cracking are two major problems with masonry structure. Both necessitate extreme care in detailing, material quality control, and field inspection during construction.

Words and Expressions

polymer	[ˈpɔlimə]	*n.* 聚合物
asphalt	[ˈæsfælt]	*n.* 沥青
ferrous	[ˈferəs]	*adj.* 铁的;含铁的
fastening	[ˈfɑːsəniŋ]	*n.* 扣紧,紧固
bearing	[ˈbeəriŋ]	*n.* 支承,支座
weldability	[ˌweldəˈbiləti]	*n.* 可焊性
corrosion	[kəˈrəuʒən]	*n.* 侵蚀,腐蚀
deformation	[ˌdiːfɔːˈmeiʃən]	*n.* 变形
ductility	[dʌkˈtiləti]	*n.* 延展性,柔软性
toughness	[ˈtʌfnis]	*n.* 韧性,刚性
deteriorate	[diˈtiəriəreit]	*v.* (使)恶化
bolt	[bəult]	*v.* 拴住
		n. 螺栓
rivet	[ˈrivit]	*v.* 铆,铆接
		n. 铆钉
yield	[jiːld]	*v.* (~ to)屈服, 屈从
compromise	[ˈkɔmprəmaiz]	*n.* 妥协,折中
sulphur	[ˈsʌlfə]	*n.* 硫黄
embrittle	[imˈbritl]	*v.* 使变脆
treat	[triːt]	*v.* 处理
addition	[əˈdiʃən]	*n.* 增加

aggregate	[ˈæɡriɡeit]	*n.* 骨料,集料
bitumen	[ˈbitjumin]	*n.* 沥青
counteract	[ˌkauntəˈrækt]	*v.* 抵消,中和
pile	[pail]	*n.* 桩
		v. 打桩,用桩支撑
masonry	[ˈmeisənri]	*n.* 砌体,砌体建筑
mortar	[ˈmɔːtə]	*n.* 灰泥,砂浆
adhesive	[ədˈhiːsiv]	*adj.* 带黏性的
bracing	[ˈbreisiŋ]	*n.* 支撑
cast iron		铸铁
wrought iron		熟铁,锻铁
rolled-steel		轧制钢材
tensile strength		抗拉强度
shock loading		冲击荷载
plain carbon steel		碳素钢
mild steel		软钢,低碳钢
reinforced concrete		钢筋混凝土
hot rolled		热轧
hard drawn		冷拉
cement paste		水泥浆
sewage treatment		污水处理
off-shore structure		海上结构
impact strength		冲击强度
cement-lime mortar		水泥石灰砂浆

Notes

① The importance of metals as constructional materials is almost invariably related to their load bearing capacity in either tension or compression and their ability to withstand limited deformation without fracture.

…as constructional materials…,副词短语修饰 metals,意思为:金属作为建筑材料的重要性。…their load bearing capacity in either tension or compression and their ability to withstand limited deformation without fracture 为名词短语作并列宾语,意思是:受拉或受压条件下的承载能力、能承受有限变形而不破裂的能力。

② The bond force is purely mechanical and arises from surface roughness and friction of the reinforcement.

arise from 意为"来源于",全句的意思为:黏结力完全是机械力,来源于表面粗糙度以及(混凝土与)钢筋的摩擦力。

③ Part of the addition forms a solid solution with the aluminum and part forms a

compound with the aluminum or with one of the other allaying elements.
由 and 引导两并列句，谓语动词均为 form，solid solution 意为固溶体。

④ The impact strength, as well as the tensile strength, of normal concrete is low and this can be improved by the introduction of randomly orientated fibers into the concrete.
as well as 引导 the tensile strength 和 the impact strength 两个并列主语，the introduction of randomly orientated fibers into the concrete 是名词短语作介词宾语，意思为：在混凝土中引入任意方向的纤维。

Exercises

Ⅰ. Translate the following words into Chinese.

1. reinforced concrete ______
2. rolled-steel ______
3. weldability ______
4. aggregate ______
5. mortar ______
6. deformation ______

Ⅱ. Translate the following words into English.

1. 冲击荷载______
2. 软钢，低碳钢______
3. 冷拉______
4. 黏土砖______
5. 收缩______

Ⅲ. Translate the following sentences into Chinese.

1. The modern civil engineer needs to deal with a variety of materials that are often integrated in the same structure, such as steel and concrete, or are used separately for construction projects, such as pavements from asphalt and Portland cement concrete.
2. The useful engineering properties of both unalloyed and alloyed aluminum are low specific gravity, resistance to corrosion, high electrical conductivity and excellent forming properties.
3. Concrete is a man-made composite, the major constituent of which is natural aggregate (such as gravel and sand) and binding medium (such as cement paste, bitumen and polymers).
4. Reinforcing techniques have been developed in recent years to extend the structural possibilities of masonry, especially improved its resistance to earthquakes.

Ⅳ. Translate the following sentences into English.

1. 钢结构的设计主要基于钢材的强度，但是钢材的延性、韧性及可焊性也是重要的特性。
2. 钢材的可焊性非常重要，因为与栓接结构和铆接结构相比，焊接结构节省重量且易于装配。
3. 普通混凝土的抗拉强度比较低，在结构应用中常规的做法或是在混凝土中放入钢筋一起抵抗拉力(钢筋混凝土)，或是用混凝土中的压力来抵消拉力(预应力混凝土或后应力混凝土)。
4. 建造过程中最大的不同是一般不需要等量的临时模板和支撑，而在混凝土结构中需要这两者。
5. 砂浆的收缩与热膨胀裂缝是砌体结构的两个主要问题。

Ⅴ. Answer the following questions briefly.

1. Please list the properties of metals that make them unique among constructional materials.
2. What are the differences between plain carbon steel and low alloy steel?
3. What are the major constituents of concrete?

Reading Materials

Basic Physical Properties of Civil Engineering Materials

Because actually different materials carry the loads, protect or decorate the structural materials, much attention is paid to study their physical properties: tensile and compressive strengths, elasticity, *ductility*(延性), and *durability*(耐久性). Among those physical properties, three important properties we will discuss below.

Tension and compression

The materials used in civil engineering projects must be able to carry loads or imposed deformation. As a matter of fact, members of a structure realize their function through tension and compression of building materials. Some examples are showed in Fig. 7-1. For the column and arch, building materials are under compression. Fibers in the upper portion of the beam are compressed and shortened, whereas those in the lower portion are tensioned and elongated. Different building materials have different abilities to carry compression and tension. For example, usually the steel is good at supporting tension and brick has excellent ability to receive compression.

***Elasticity* (弹性) and *plasticity* (塑性)**

Materials for civil engineering must be able to carry loads, or weight, without changing shape permanently. When a load is applied to a structure member, it will *deform*(变形), that is, a wire will stretch or a beam will bend. However, when the load is removed, the wire and the beam come back to the original positions. This material property is called elasticity. If a material is not elastic and a deformation is present in the structure after removal of the load, this material property is called plasticity.

All materials used in civil engineering structures, such as stone and brick, wood, steel, aluminum, reinforced concrete, and plastics, behave elastically within a certain defined range of loading. If the loading is increased above the range, two types of behavior can occur: brittle and plastic. In the former, the material will break suddenly. In the latter, the material begins to flow at a certain load (*yield strength* 屈服应力), ultimately leading to fracture. As examples, steel exhibits plastic behavior, and stone is brittle. The *ultimate strength*(极限强度) of a material is measured by the stress at which failure (fracture) occurs.

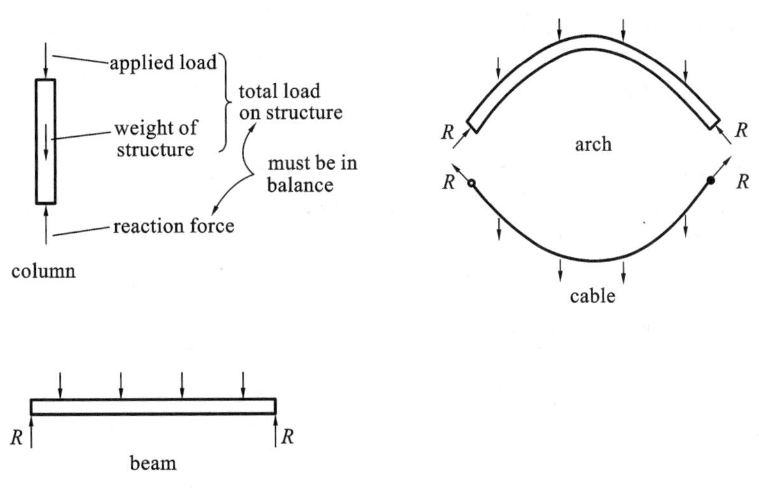

Fig. 7-1 Column, beam, arch, cable carrying loads

***Stiffness*(刚度)**

This property is defined by the *elastic modulus*(弹性模量)(E), which is the ratio of the stress (force per unit area), to the strain (deformation per unit length): $\frac{\text{stress}}{\text{strain}}=E$. The elastic modulus, therefore, is a measure of the resistance of a material to deformation under load. For two materials of equal area under the same load, the one with the higher elastic modulus has smaller deformation. Structural steel, which has an elastic modulus 2 100 000 kilograms per square centimeter, is 3 times as *stiff*(刚性的,硬的) as aluminum, 10 times as stiff as concrete, and 15 times as stiff as wood.

Reading Skills(10)—Stems and Affixes(Ⅱ)

In the chart below, some of the most common suffixes are listed alphabetically. The meaning of each suffix is given as an area of meaning because most often there is no one single specific meaning. In the right-hand column, space is left for you to record examples of words with the suffix. Your dictionary will provide examples. Choose those which are familiar or *potentially*(潜在的) useful to you.

Suffix	Area of meaning	Examples
-able,-ible	capable of being	
-ance,-ence	state, condition, or quality	
-ary, etc.	a place of	
-ate	relating to	
-ation,-tion	condition or the act of, quality of	
-cy	state of	
-dom	state, condition	
-ee	the object or receiver of action	
-en	to make, of the nature of, small	
-er	the one who	
-et	small	
-ful	full of, or characterized by	
-fy	to make	
-hood	state or duality of	
-ic,-ical	pertaining to	
-ious,-ous	full of, of the nature of	
-ize,-ise	to make like or affect with, cause to become	
-ish	to form adjectives from nouns; belonging to, like	
-ism,-ist,-istic	action or practice, state or condition	
-ive	of the nature of	
-ity,-ty	condition, state of	
-less	without, loose from	
-let	small	
-ly	like	
-ness	state, conditions, quality	
-ship	condition, skill, state, character, office	

Reading Skills(11)—Prefixes(Ⅰ)

re-(again)　un-(not)　im-(not)

A knowledge of these *prefixes*(前缀) will provide you with a key to the meaning of many unfamiliar words. Some words that have these prefixes are:

reprint, reuse, rewrite, remove, replace, unknown, uneconomical, unhappy, uncomfortable, unreal, unpleasant; impossible, impassable, impure, impatient, imperfect, improper.

Sometimes re-, un-, and im- are not prefixes, they do not join with a base word to change its meaning. In the words rest, unit, re-, un-, and im-are not prefixes.

Directions: All the words begin with re-, un-, or im-. If the beginning is a prefix, write P. If it is not a prefix, write N.

1. university ________　2. ready ________　3. immaterial ________
4. imagine ________　5. under ________　6. reward ________
7. unable ________　8. reason ________　9. united ________
10. immediate ________

Lesson 8

FORCE METHOD OF STRUCTURAL ANALYSIS

Force method of analysis is one of the basic methods of analysis of structures. It is proposed to outline the primary procedure in this paper.

Description of method

First of all, the degree of statical indeterminacy is determined. A number of releases equal to the degree of indeterminacy is now introduced, each release being made by the removal of an external or an internal force. The releases must be chosen so that the remaining structure is stable and statically determinate. However in some cases the number of releases can be less than the degree of indeterminacy, provided the remaining statically indeterminate structure is so simple that it can be readily analyzed. ① In all cases, the released forces, which are also called redundant forces, should be carefully chosen so that the released structure is easy to analyze.

The releases introduce inconsistencies in displacements, and as a second step these inconsistencies or "errors" in the released structure are determined. In other words, we calculate the magnitude of the "errors" in the displacements corresponding to the redundant forces. These displacements may be due to external applied loads, settlement of supports, or temperature variation.

The third step consists of a determination of the displacements in the released structure due to unit values of the redundancies(cf. Fig. 8-1(d) and (e)). These displacements are required at the same location and in the same direction as the error in displacements determined in step 2.

The values of the redundant forces necessary to eliminate the errors in the displacements are now determined. This requires the writing of superposition equations in which the effects of the separate redundancies are added to the displacements of the released structure.

Hence, we find the forces on the original indeterminate structure: they are the sum of the correction forces (redundancies) and forces on the released structure.

This brief description of the application of the force method will now be illustrated by an example.

Example 8-1

Fig. 8-1(a) shows a beam *ABC* fixed at *C*, resting on roller supports at *A* and *B*, and carrying a uniform load of q per unit length. The beam has a constant flexural rigidity *EI*. Find the reactions of the beam.

The structure is statically indeterminate to the second degree, so that two redundant forces have to be removed. Several choices are possible, e. g. , the moment and the vertical reaction at C, or the vertical reactions at A and B. For the purposes of this example, we shall remove the vertical reaction at B and the moment at C. The released structure is then a simple beam AC with redundant forces and displacements as shown in Fig. 8-1(b). The location and direction of the various redundancies and displacements is referred to as a coordinate system.

The positive directions of the redundancies F_1 and F_2 are chosen arbitrarily but the positive directions of the displacements at the same location must always accord with those of the redundancies. ② The arrows in Fig. 8-1(b) indicate the chosen positive directions in the present case and, since the arrows indicate forces as well as displacements, it is convenient in a general case to label the coordinates by numerals 1, 2, …, n. ③

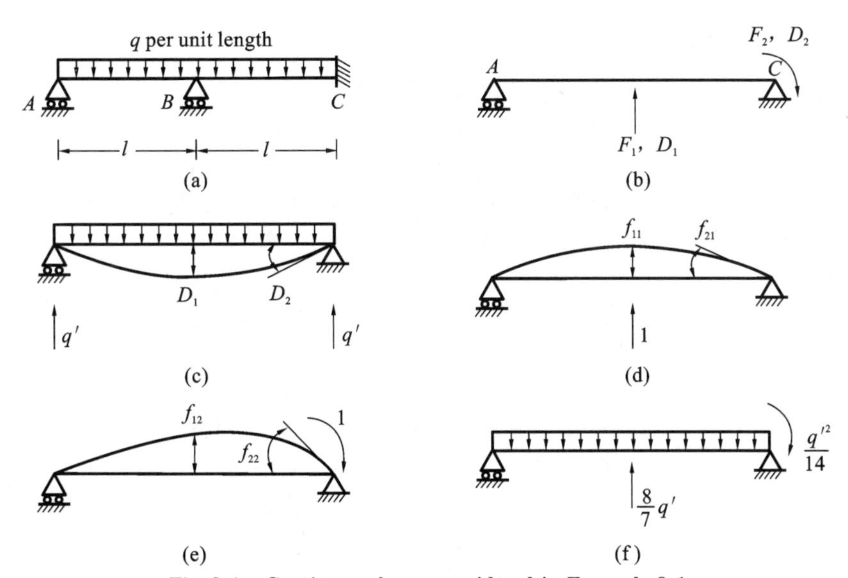

Fig. 8-1 Continuous beam considered in Example 8-1

(a) statically indeterminate beam; (b) coordinate system;

(c) external load on released structure; (d) $F_1=1$; (e) $F_2=1$; (f) redundancies

Following this system, Fig. 8-1(c) shows the displacements at B and C as D_1 and D_2 respectively. In fact, as shown in Fig. 8-1(a), the actual displacements at those points are 0, so that D_1 and D_2 represent the inconsistencies in deformation.

The magnitude of D_1 and D_2 can be calculated from the behavior of the simply supported beam of Fig. 8-1(c). For the present purposes, we can calculate that

$$D_1 = -\frac{5ql^4}{24EI} \quad \text{and} \quad D_2 = -\frac{ql^3}{3EI} \qquad 8\text{-}1$$

The negative signs show that the displacements are in directions opposite to the positive directions chosen in Fig. 8-1(b). ④

The displacements due to unit values of the redundancies are shown in Figs. 8-1(d) and(e). Their displacements are as follows

$$\left.\begin{array}{ll} f_{11}=\dfrac{l^3}{6EI} & f_{12}=\dfrac{l^2}{4EI} \\ f_{21}=\dfrac{l^2}{4EI} & f_{22}=\dfrac{2l}{3EI} \end{array}\right\} \quad 8\text{-}2$$

The general coefficient f_{ij} represents the displacement at the coordinate i due to a unit redundant at the coordinate j.

The geometry relations express the fact that the final vertical translation at B and the rotation at C vanish. The final displacements are the result of the superposition of the effect of the external loading and of the redundants on the redundancies structure. Thus the geometry-relations can be expressed as

$$\left.\begin{array}{l} D_1+f_{11}F_1+f_{12}F_2=0 \\ D_2+f_{21}F_1+f_{22}F_2=0 \end{array}\right\} \quad 8\text{-}3$$

Words and Expressions

release	[ri'liːs]	*n.* 释放,发行
displacement	[dis'pleismənt]	*n.* 位移,排水量
error	['erə(r)]	*n.* 误差,偏差
cf.		*n.* 比较(=confer)
eliminate	[i'limineit]	*v.* 排除,消除
superposition	[ˌsjuːpəpə'ziʃən]	*n.* 叠加,重叠
moment	['məumənt]	*n.* 力矩
coefficient	[ˌkəui'fiʃənt]	*n.* [数]系数
rotation	[rəu'teiʃən]	*n.* 旋转
force method		(结构力学中的)力法
structural analysis		结构分析
degree of statical indeterminacy		超静定次数
statical indeterminacy		超静定
statically determinate		静定的
redundant force		赘余力
released structure		放松结构,去掉多余约束的结构
uniform load		均布荷载
flexural rigidity		抗挠刚度,弯曲刚度
coordinate system		坐标系
continuous beam		连续梁
simply supported beam		简支梁

Notes

① However in some cases the number of releases can be less than the degree of indeterminacy, provided the remaining statically indeterminate structure is so simple that it can be readily analyzed.
然而，在一些情况下，如果所剩的静定结构简单到很容易分析，去掉的约束的个数可以少于超静定次数。

② The positive directions of the redundancies F_1 and F_2 are chosen arbitrarily but the positive directions of the displacements at the same location must always accord with those of the redundancies.
外力 F_1 和 F_2 的正向是任意确定的，但是相同位置位移正向的确定要与赘余力相一致。

③ The arrows in Fig. 8-1(b) indicate the chosen positive directions in the present case and, since the arrows indicate forces as well as displacements, it is convenient in a general case to label the coordinates by numerals 1, 2, …, n.
图 8-1(b)中的箭头指出了这种情况下的正向(箭头既可以代表力的方向，同时也可以指示位移的方向)，一般情况下以 1,2,…,n 来表示坐标轴比较方便。

④ The negative signs show that the displacements are in directions opposite to the positive directions chosen in Fig. 8-1(b).
这里负号表示位移与图 8-1(b)中所选的正向相反。

Exercises

Ⅰ. Translate the following words into Chinese.

1. inconsistency ____________________
2. coefficient ____________________
3. redundant ____________________
4. statical indeterminacy ____________________
5. constant flexural rigidity ____________________
6. simply supported beam ____________________
7. geometry-relations ____________________

Ⅱ. Translate the following words into English.

1. 力矩 ____________________
2. 均布荷载 ____________________
3. 等价体系 ____________________
4. 连续梁 ____________________
5. 二次超静定 ____________________
6. 放松结构 ____________________
7. 滚轴支座 ____________________

Ⅲ. Translate the following sentences into Chinese.

1. A number of releases equal to the degree of indeterminacy is now introduced, each release being made by the removal of an external or an internal force.
2. These displacements are required at the same location and in the same direction as the error in displacements determined in step 2.
3. The releases introduce inconsistencies in displacements, and as a second step these inconsistencies or "errors" in the released structure are determined. In other words, we calculate the magnitude of the "errors" in the displacements corresponding to the redundant forces.
4. The structure is statically indeterminate to the second degree, so that two redundant forces have to be removed.

Ⅳ. Translate the following sentences into English.

1. 这些位移可能是由于外加荷载、基础沉降或者温度变化引起的。
2. 因此,我们找到作用于最初的超静定结构上的力:它们是赘余力和作用在基本结构上的力的和。
3. 最终位移是基本结构上附加荷载和外荷载的效应叠加产生的。

Ⅴ. Answer the following questions briefly.

1. Please describe the force method of analysis.
2. What does the general coefficient f_{ij} represent?
3. What is the geometry-relation?

Reading Materials

Finite Element Method

The *finite element method*(有限元法) for analyzing structural parts has now been around for about 30 years, but although it is generally accepted as an extremely valuable tool, many engineers do not know how to go about using it and very few engineers understand it. One of the main reasons for this is that the subject has generally been surrounded by a high level of research activity. Coupled with this is the fact that because of the amount of calculations which the method involves, it tended to go directly from its *embryonic*(初期的) stage to an advanced computing stage. Basically the finite element method involves the application of the three basic conditions, i.e. *equilibrium of forces*(力的平衡), *compatibility of displacement*(位移协调) and *stress-strain relationships*(应力-应变关系). It should also be pointed out that due to the versatility of the finite element method it can be applied to very complex problems and the matrix methods involved in the solution of these can provide material for a whole textbook in their own right. It is proposed to provide a simple introduction to the finite element method and remove some of the mystery which surrounds the subject by illustrating its application to a number of typical problems.

If a truss of the type shown in Fig. 8-1(a) is being analyzed then it is a straightforward exercise because it is formed from *discrete*(离散的) members. The paths of the force transmission through the truss to the ground are readily apparent and the forces may be determined using equilibrium of individual element.

If the truss had been statically indeterminate then both equilibrium equations and deformation compatibility would be necessary, but the method of solution is well established. If, however, a plate of the same shape as the truss (Fig. 8-2(b)) is to be analyzed then this is not so straightforward. The reason is that the plate is an elastic *continuum*(连续体) and since the force transmission paths are not readily apparent the problem does not lend itself to simple mathematical analysis. Although, for the purpose of analysis, it might be tempting to consider the truss as being equivalent to the plate, this would not give accurate results since it ignores the restraining effect which all points in a continuum will experience and exert on neighboring points. However, if the continuum was considered to be subdivided into a large number of triangular panels (see Fig. 8-2(c)), it should be possible to develop a picture of the stress distribution in the whole plate by analyzing each of the small panels in turn. To do this it would of course be necessary to (a) analyze the equilibrium of each of the triangular panels in relation to its neighbors and (b) have available equations for the *geometry of deformation*(几何变形) and the stress-strain relationships for a triangular panel. This subdivision of a continuum into large number of discrete element is the basis of the finite element method of stress analysis. The triangular panels referred to in this example are the "elements", but this is only one type rectangular element. Others include a spring element (one-dimensional) and solid elements (three-dimensional) as shown in Fig. 8-3.

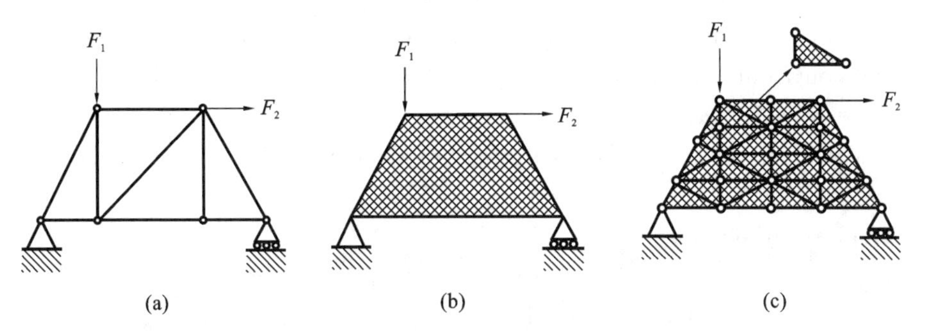

Fig. 8-2 Truss and plate model

The accuracy of the solution depends on the number of subdivisions (elements); the more there are, the greater the accuracy is. However, although the analysis of each individual element is straightforward, the analysis of a large number of elements becomes extremely tedious. For this reason finite element solutions to problems are generally carried out on computers and there are many commercial software packages available. To some extent this has led to the current situation where many engineers are put off by apparent complexity of the subject and they leave it to the experts who tend to attach a certain mystique to the subject through the use of computer.

As an introduction to finite element methods, it is convenient to consider the matrix analysis of skeletal structures using the stiffness method.

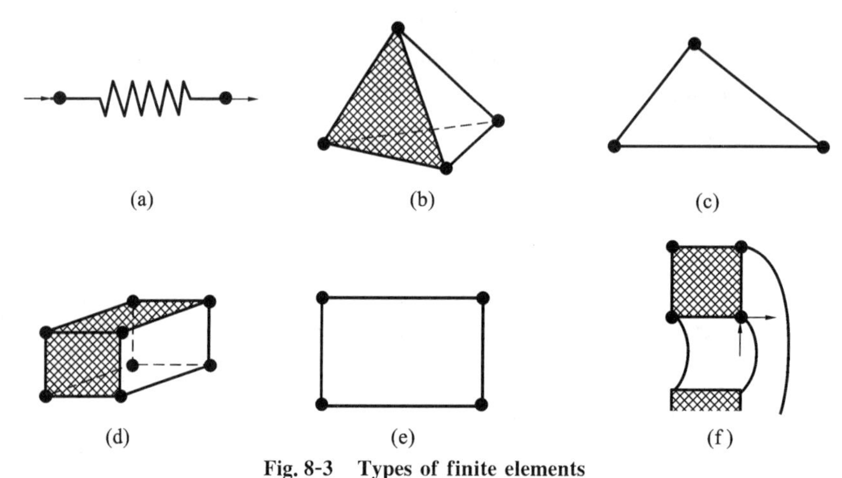

Fig. 8-3 Types of finite elements

(a)spring; (b)tetrahedron; (c)triangle; (d)hexahedron; (e)axisymmetric; (f)quadrilateral

Reading Skills(12)—Prefixes(Ⅱ)

The prefixes de-, ex- and ab- have many meanings.

de-=away, from, down, off, very, wholly, reverse, of, undo

ex-=out, out of, forth, from, beyond, away, upward, without

ab-=away, from, down

Add de-, ex- or ab- to each word part to make a word that will fit the *definition*(定义) given.

Write the whole word.

1. ________ pression (low spirits)
2. ________ tend (reach beyond)
3. ________ sent (not present; missing)
4. ________ for (put off)
5. ________ sorb (to take or suck in)
6. ________ stain (do without, refrain)

7. ________ serve (be worthy of)
8. ________ stract (not real or solid)
9. ________ use (use badly)
10. ________ port (sent out of)
11. ________ part (go away)
12. ________ ception (something outside the rule)
13. ________ cuse (free from)
14. ________ normal (different from; unusual)

The prefix sub-, means under, below, beneath, lower than, or less than. It can be combined with many base words to add these meanings. It is not always easy to recognize, because its spelling often changes according to the word with which it is combined. It usually becomes:

suc- before c
suf- before f
sug- before g
sum- before m
sup- before p
sur- before r

Decide which form of the prefix sub- goes with each word part below. Add the prefix and write the word.

1. ________ pose　　2. ________ stance　　3. ________ ceed
4. ________ fix　　5. ________ round　　6. ________ gest
7. ________ fer　　8. ________ mon　　9. ________ port
10. ________ ject

Reading Skills(13)—Suffixes(Ⅰ)

-ist (one who practices or believes in)
-ism (practice of or belief in)
-istic (pertaining to practice of or belief in)

The suffixes above may be used with many English words, including

individual　　real
terror　　national
ideal

Combine each word above with the suffixes -ist,-ism,-istic to form a word that best fits each sentence.

1. One who sees things as they actually are is a ________________.
2. People who are *excessively*(过分地) devoted to their country can be described as ________________.

3. The ________________ believes in personal freedom and personal enterprise.
4. Behavior or thought based on things as they should be is ________________ behavior or thought.
5. One who views things as they actually are is facing life with ____________.
6. A ________________ is one who believes in using *intimidation*(胁迫) as a weapon.
7. The ________________ is often regarded as an extremely *impractical*(不切实际的) person.
8. A person who makes his own rules is extremely ________.
9. A ________________ is more interested in his own country than the others.
10. Acting on high principles shows that one has ________________.

Lesson 9

PLASTIC HINGES AND COLLAPSE MECHANISMS

If a short segment of a reinforced concrete beam is subjected to a bending moment, curvature of the beam axis will result, and there will be a corresponding rotation of one face of the segment with respect to the other. It is convenient to express this in terms of an angular change per unit length of the member. The relation between moment and angle change per unit length of beam, or curvature, at a reinforced concrete beam section subject to tensile cracking was developed in previous papers. Methods were presented there by which the theoretical moment-curvature graph might be drawn for a given beam cross section, as in Fig. 9-1.

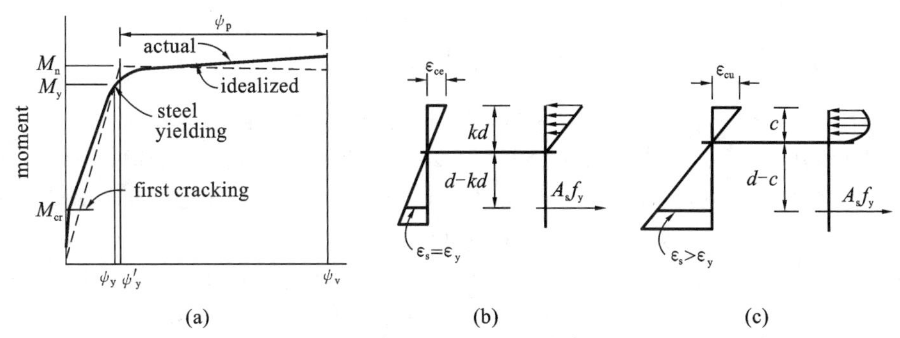

Fig. 9-1 Plastic hinge characteristics in a reinforced concrete member

(a) typical moment-rotation diagram; (b) strains and stresses at start of yielding; (c) strains and stresses at incipient failure

The actual moment-curvature relationship measured in beam tests differs somewhat from that shown in Fig. 9-1, mainly because, from tests, curvatures are calculated from average strains measured over a finite gage length, usually about equal to the effective depth of the beam. In particular, the sharp increase in curvature upon concrete cracking shown in Fig. 9-1 is not often seen, because the crack occurs at only one discrete location along the gage length. Elsewhere, the uncracked concrete shares in resisting flexural tension, resulting in what is known as tension stiffening.[①] This tends to reduce curvature. Furthermore, the exact shape of the moment-curvature relation depends strongly upon the steel ratio as well as upon the exact stress-strain curves for the concrete and steel.

Fig. 9-1 shows a somewhat simplified moment-curvature diagram for all actual concrete beam section having tensile steel ratio of about one-half the balanced value. The diagram is linear up to the cracking moment M_{cr}, after which a nearly straight line of somewhat flatter slope is obtained. At the moment that initiates yielding, M_y, the unit rotation starts to increase disproportionately. Further increase in applied moment causes extensive inelastic rotation until eventually the compressive strain limit of the concrete is reached at the ultimate rotation ψ_u. The resisting moment at ultimate capacity is often somewhat above the calculated flexural strength M_n, due largely to strain hardening of the reinforcement.

Unit rotation

The effect of inelastic concrete response prior to steel yielding is small for typically under-reinforced sections, as is indicated in Fig. 9-1, and the yield moment can be calculated based on the elastic concrete stress distribution shown in Fig. 9-1(b):

$$M_y = A_s f_y \left(d - \frac{kd}{3}\right) \quad 9\text{-}1$$

where kd is the distance from the compression face to the cracked elastic neutral axis. The ultimate moment capacity M_n, based on Fig. 9-1(c), is calculated by the usual expression

$$M_n = A_s f_y \left(d - \frac{a}{2}\right) = A_s f_y \left(d - \frac{\beta_1 c}{2}\right) \quad 9\text{-}2$$

For purposes of limit analysis, the M-ψ curve is usually idealized, as shown by the dashed line in Fig. 9-1(a). The slope of the elastic portion of the curve is obtained with satisfactory accuracy using the moment of inertia of the cracked transformed section. After the calculated ultimate moment M_n is reached, continued plastic rotation is assumed to occur with no change in applied moment. The elastic curve of the beam will show an abrupt change in slope at such a section. The beam behaves as if there were a hinge at that point. However, the hinge will not be "friction-free", but will have a constant resistance to rotation.

If such a plastic hinge forms in a determinate structure, as shown in Fig. 9-2, uncontrolled deflection takes place, and the structure will collapse. The resulting system is referred to as a mechanism, an analogy to linkage systems in mechanics. In general, one can say that a statically determinate system requires the formation of only one plastic hinge in order to become a mechanism.

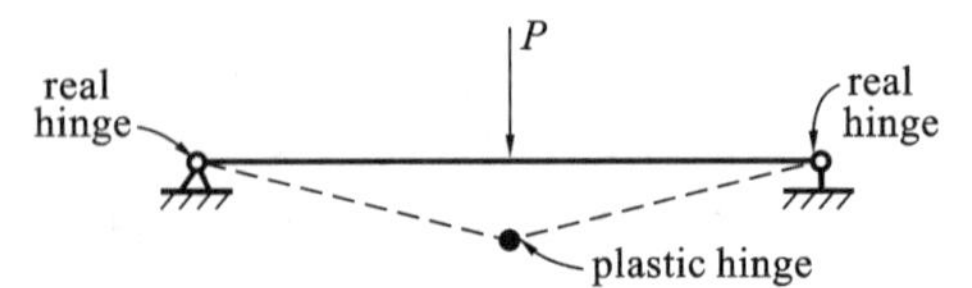

Fig. 9-2 Statically determinate system after the formation of plastic hinges

This is not so for indeterminate structures. In this case, stability may be maintained even though hinges have formed at several cross sections. The formation of such hinges in indeterminate structures permits a redistribution of moments within the beam or frame. It will be assumed for simplicity that the indeterminate beam of Fig. 9-3(a) is symmetrically reinforced, so that the negative bending capacity is the same as the positive. Let the load P be increased gradually until the elastic moment at the fixed support, $\frac{3}{16}PL$, is just equal to the plastic moment capacity of the section M_n. This load is

$$P=P_{el}=\frac{16}{3}\frac{M_n}{L}=5.33\frac{M_n}{L} \qquad 9\text{-}3$$

At this load, the positive moment under the load is $\frac{5}{32}PL$, as shown in Fig. 9-3(b). The beam still responds elastically everywhere but at the left support. At that point the actual fixed support can be replaced for purposes of analysis with a plastic hinge offering a known resisting moment M_n. Because a redundant reaction has been replaced by a known moment, the beam is now determinate. ②

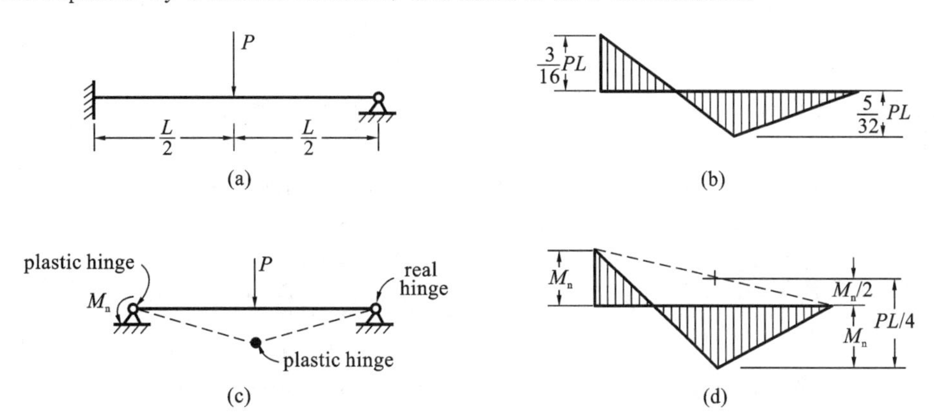

Fig. 9-3 Indeterminate beam with plastic hinges at support and midspan

The load can be increased further until the moment under the load also becomes equal to M_n, at which load the second hinge forms. The structure is converted into a mechanism, as shown in Fig. 9-3(c), and collapse occurs. The moment diagram at collapse load is shown in Fig. 9-3(d).

The magnitude of load causing collapse is easily calculated from the geometry of Fig. 9-3(d)

$$M_n+\frac{M_n}{2}=\frac{PL}{4} \qquad 9\text{-}4$$

from which

$$P=P_n=\frac{6M_n}{L} \qquad 9\text{-}5$$

By comparison of Eq. 9-4 and Eq. 9-5, it is evident that an increase in P of 12.5 percent is possible, beyond that load which caused the formation of the first plastic hinge, before the beam will actually collapse. ③ Due to the formation of plastic hinges, a redistribution of moments has occurred such that, at failure, the ratio between the positive moment and negative moment is equal to that assumed in reinforcing the structure.

Words and Expressions

segment	[ˈsegmənt]	*n*. (分割的)部分,片段
angular	[ˈængjulə]	*adj*. 有角的,成角(度)的
incipient	[inˈsipiənt]	*adj*. 开始的,初期的
discrete	[diˈskriːt]	*adj*. 不连续的,离散的
disproportionate	[ˌdisprəˈpɔːʃənət]	*adj*. 不成比例的,不均衡的
inertia	[iˈnəːʃjə]	*n*. 惯性,惯量
abrupt	[əˈbrʌpt]	*adj*. 突然的,急剧的
redundant	[riˈdʌndənt]	*adj*. 冗余的,超静定的
gage length		测量标距
effective depth		有效高度
strain hardening		应变硬化(强化)
under-reinforced		配筋适量的
ultimate moment capacity		极限受弯承载力
transformed section		换算截面
moment diagram		弯矩图

Notes

① In particular, the sharp increase in curvature upon concrete cracking shown in Fig. 9-1 is not often seen, because the crack occurs at only one discrete location along the gage length. Elsewhere, the uncracked concrete shares in resisting flexural tension, resulting in what is known as tension stiffening.
尤其是,由于裂缝通常只出现在沿应变计长度的某个随机位置,因此在图 9-1 中由于混凝土开裂而导致的曲率的迅速增长并不常见;其他地方,未开裂的混凝土参与抵抗弯曲拉伸,即产生了拉伸强化。

② At that point the actual fixed support can be replaced for purposes of analysis with a plastic hinge offering a known resisting moment M_n. Because a redundant reaction has been replaced by a known moment, the beam is now determinate.
出于分析目的,该点的实际固端支承可以用一个提供已知抵抗弯矩 M_n 的塑性铰替代。由于一个多余反力被已知弯矩所替代,梁就变为静定的了。

③ By comparison of Eq. 9-4 and Eq. 9-5, it is evident that an increase in P of 12.5 percent is possible, beyond that load which caused the formation of the first plastic hinge, before the beam will actually collapse.
比较公式 9-4 和公式 9-5,显然在梁被实际破坏之前,P 在引起第一个塑性铰形成的荷载的基础上再增加 12.5%是可能的。

Exercises

Ⅰ. Translate the following words into Chinese.

1. plastic hinge ________________
2. angular change per unit length ________________
3. steel ratio ________________
4. cracking moment ________________
5. moment of inertia ________________
6. tension stiffening ________________

Ⅱ. Translate the following words into English.

1. 弯矩-曲率关系________________
2. 应变计长度范围内测得的平均应变________________
3. 应变硬化________________
4. 极限压应变________________
5. 抗弯承载力________________
6. 梁的有效高度________________
7. 弯矩图________________
8. 界限配筋率________________

Ⅲ. Translate the following sentences into Chinese.

1. The actual moment-curvature relationship measured in beam tests differs somewhat from that shown in Fig. 9-1, mainly because, from tests, curvatures are calculated from average strains measured over a finite gage length, usually about equal to the effective depth of the beam.
2. The effect of inelastic concrete response prior to steel yielding is small for typically under-reinforced sections, as is indicated in Fig. 9-1, and the yield moment can be calculated based on the elastic concrete stress distribution shown in Fig. 9-1(b).
3. After the calculated ultimate moment M_n is reached, continued plastic rotation is assumed to occur with no change in applied moment. The elastic curve of the beam will show an abrupt change in slope at such a section. The beam behaves as if there were a hinge at that point. However, the hinge will not be "friction-free," but will have a constant resistance to rotation.
4. The resulting system is referred to as a mechanism, an analogy to linkage systems in mechanics.

Ⅳ. Translate the following sentences into English.

1. 在最大弯矩等于塑性弯矩时，其曲率变得非常大，并出现塑性流动。
2. 塑性铰总是在弯矩最大的截面上出现，它的出现意味着构件将沿铰截面转动。

3. 塑性铰的形成将使构件或结构成为破坏结构。
4. 铰中塑性区的长度可以较容易地计算出来。

Ⅴ. Answer the following questions briefly.

1. By what factors is the shape of moment-curvature mainly decided?
2. Describe the general development of moment-curvature curve, and pay attention to the key points.
3. What are the main characteristics of plastic hinges?
4. How does plastic hinge work in determinate and indeterminate structure?
5. What is supposed to be done with a redundant reaction in analysis?

Reading Materials

Limit Analysis

Presently, most reinforced concrete structures are designed for moments, shears, and thrusts found by elastic theory. On the other hand, the actual proportioning of members is done by strength methods, with the recognition that inelastic section and member response would result upon overloading. *Factored loads*(乘上系数的荷载)are used in the elastic analysis to find moments in a continuous beam, for example, after which the critical beam sections are designed with the knowledge that the steel would be well into the yield range and the concrete stress distribution very nonlinear before final collapse. Clearly this is an inconsistent approach to the total analysis-design process, although it can be shown to be both safe and conservative. A beam or frame so analyzed and designed will not fail at load lower than the value calculated in this way.

On the other hand, it is known that a continuous beam or frame normally will not fail when the ultimate moment capacity of just one critical section is reached. A *plastic hinge*(塑性铰) will form at that section, permitting large rotation to occur at essentially constant resisting moment and thus transferring load to other locations along the span where the limiting resistance has not yet been reached. Normally in a continuous beam or frame, excess capacity will exist at those other locations, because they would have been reinforced for moments resulting from different load distributions selected to produce maximum moments at those other locations.

As loading is further increased, additional plastic hinges may form at other locations along the span and eventually result in collapse of the structure, but only after a significant *redistribution of moments*(弯矩重分布) has occurred. The ratio of negative to positive moments found from elastic analysis is no longer correct, for example, and the true ratio after redistribution depends upon the flexural strengths actually provided at the hinging sections.

Recognition of redistribution of moments can be important, because it permits a more realistic appraisal of the actual load-carrying capacity of a structure, thus leading to improved economy. In addition, it permits the designer to modify, within limits, the moment diagrams for which members are to be designed. Certain sections can be deliberately under-reinforced if moment resistance at adjacent critical sections is increased correspondingly. Adjustment of design moments in this way enables the designer to reduce the congestion of reinforcement that often occurs in high-moment areas, such as at the beam-column joints.

The formation of plastic hinges is well established by tests, which were carried out in the George Winter Laboratory at Cornell University. The three-span continuous beam illustrates the inelastic response typical of heavily overloaded members. It was reinforced in such a way that plastic hinges would form at the interior support sections before the limit capacity of the sections elsewhere was reached. The beam continued to carry increasing load well beyond the load that produced first yielding at the supports. The extreme deflections and sharp changes in slope of the member axis that are seen here were obtained only slightly before final collapse.

The *inconsistence*(不协调) of the present approach to the total analysis-design process, the possibility of utilizing the *reserve strength*(储备强度)of concrete structures resulting from moment redistribution, and the opportunity to reduce steel congestion in critical regions have motivated considerable interest in limit analysis for reinforced concrete based on the concepts just described. For beams and frames, ACI code 8.4 permits redistribution of moments, depending upon the tensile steel ratio. For slabs, which generally use very low steel ratios and consequently have great *ductility*(延性), plastic design methods are especially suitable.

Reading Skills(14)—Suffixes(Ⅱ)

The suffixes -ary,-arium,-ory, and -orium all mean "a place for". The word in column Ⅱ have the suffix -ary,-arium,-ory, or -orium.

Write the word that fits each definition in column Ⅰ.

Ⅰ	Ⅱ
1. place for sunbathing	auditorium
2. place for observation, often astronomical	granary
3. place where aquatic animals are displayed	laboratory
4. place for listening	planetarium
5. place where books are kept	aquarium
6. place for work, usually scientific	conservatory
7. place where criminals are confined	solarium

8. room enclosed in glass, for preserving growing plants	observatory
9. place where grain is stored	penitentiary
10. place where models of the solar system are displayed	library

Reading Skills(15)—Suffixes(Ⅲ)

The suffixes -fy and -en both mean "to make". Usually only one of these suffixes can be used with a particular word. Decide which suffix goes with each word part.

Add the suffix and write the word.

quick ________	ampli ________
simpli ________	intensi ________
deep ________	height ________
short ________	dark ________
electri ________	lique ________

Lesson 10

REINFORCEMENT IN RECTANGULAR RC BEAMS

Before proceeding with the derivation of beam expressions, it is necessary to define certain terms relating to the amount of tensile steel used in a beam. These terms include balanced steel ratio, underreinforced beams[①], and overreinforced beams.

A beam that has a balanced steel ratio is one for which the tensile steel will theoretically start to yield and the compression concrete reach its ultimate strain at exactly the same load. Should a beam have less reinforcement than required for a balanced ratio, it is said to be underreinforced; if more, it is said to be overreinforced.

If a beam is underreinforced and the ultimate load is approached, the steel will begin to yield even though the compression concrete is still understressed. If the load is further increased, the steel will continue to elongate, resulting in appreciable deflections and large visible cracks in the tensile concrete. As a result, the users of the structure are given notice that the load must be decreased or else the result will be considerable damage or even failure. If the load is increased further, the tension cracks will become even larger and eventually the compression side of the concrete will become overstressed and fail.

If a beam should be overreinforced, the steel will not yield before failure. As the load is increased, deflections are not noticeable even though the compression concrete is highly stressed, and failure occurs suddenly without warning to the occupants. Rectangular beams will fail in compression when strains are about 0.003 to 0.004 for ordinary grades of concrete.

Obviously, overreinforcing is a situation to be avoided if at all possible, and the code, by limiting the percentage of tensile steel that may be used in a beam, ensures the design of underreinforced beams and thus the ductile type of failures that provide adequate "running time".[②]

Derivation of beam expression

Tests of reinforced concrete beams confirm that strains vary in proportion to distances from the neutral axis even on the tension sides and even near ultimate loads. Compression stresses vary approximately in a straight line until the maximum stress equals about 0.50 f'_c. This is not the case, however, after stresses go higher. When the ultimate load is reached, the strain and stress variations are approximately as shown in Fig. 10-1.

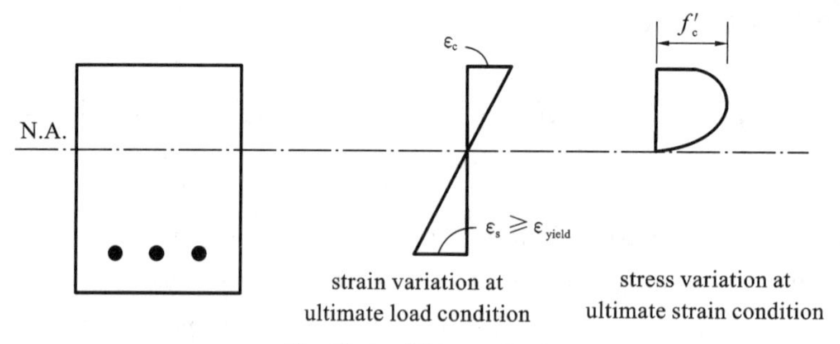

Fig. 10-1 Ultimate load

The compressive stresses vary from 0 at the neutral axis to a maximum value at or near the extreme fiber. The actual stress variation and the actual location of the neutral axis vary somewhat from beam to beam depending on such variables as the magnitude and history of past loadings, shrinkage and creep of the concrete, size and spacing of tension cracks, speed of loading, and so on. ③

If the shape of the stress diagram were the same for every beam, it would easily be possible to derive a single rational set of expressions for flexural behavior. Because of these stress variations, however, it is necessary to base the strength design upon a combination of theory and test results.

Although the actual stress distribution given in Fig. 10-2(b) may seem to be an important matter, any assumed shape can be used practically if the resulting equations compare favorably with test results. The most common shapes proposed are the rectangle, parabola, and trapezoid, with the rectangular shape used in this text as shown in Fig. 10-2(c) being the most common one.

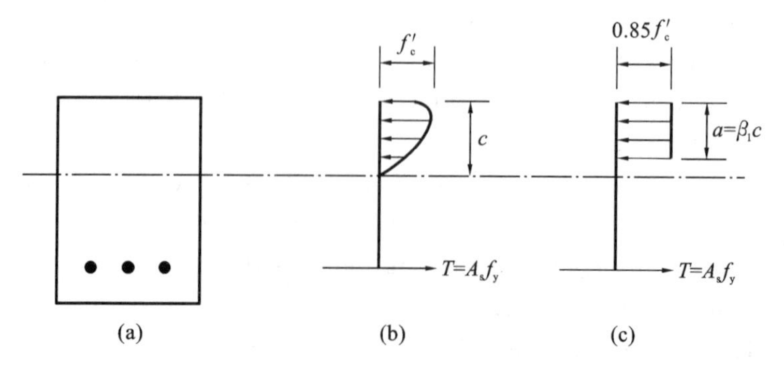

Fig. 10-2 Some possible stress distribution shapes

If the concrete is assumed to crush at a strain of about 0.003 (which is a little conservative for most concretes) and the steel to yield at f_y, it is possible to make a reasonable derivation of beam formulas without knowing the exact stress distribution. However, it is necessary to know the value of the total compression and its centroid.

Whitney replaced the curved stress block with an equivalent rectangular block of intensity $0.85 f'_c$ and depth $a=\beta_1 c$, as shown in Fig. 10-2(c). The area of this rectangular block should equal to that of the curved stress block, and the centroids of the two blocks should coincide. Sufficient test results are available for concrete beams to provide the depths of the equivalent rectangular stress blocks. The values of β_1 given by the code are intended to give this result. For f'_c values of 4 000 psi or less, $\beta_1=0.85$. In SI units, β_1 is to be taken equal to 0.85 for concrete strengths up to and including 30 MPa.

For concretes with $f'_c > 30$ MPa, β_1 can be determined with the following expression:

$$\beta_1=0.85-\left(\frac{f'_c-30}{7}\right)(0.05)\geqslant 0.65 \qquad 10\text{-}1$$

Based on these assumptions regarding the stress block, statics equations can easily be written for the sum of the horizontal forces and for the resisting moment produced by the internal couple. These expressions can then be solved separately for a and for the moment M_n.

A very clear statement should be made here regarding the term M_n because it otherwise can be confusing to the reader. M_n is defined as the theoretical or nominal resisting moment of a section. The usable strength of a member equals its theoretical strength times the strength reduction factor, or, in this case, ϕM_n. The usable flexural strength of a beam is defined as M_u.

$$M_u=\phi M_n \qquad 10\text{-}2$$

For writing the beam expressions, reference is made to Fig. 10-3. Equating the horizontal forces C and T and solving for a, we obtain: $0.85 f'_c ab=A_s f_y$

$$a=\frac{A_s f_y}{0.85 f'_c b}=\frac{\rho f_y d}{0.85 f'_c} \qquad 10\text{-}3$$

where $\rho=\frac{A_s}{bd}$ = percentage of tensile steel

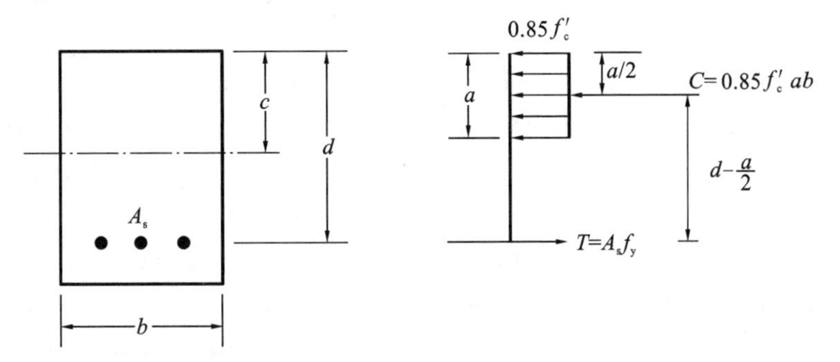

Fig. 10-3 Calculation of bearing capacity of the single steel rectangular beam section

Because the reinforcing steel is limited to an amount such that it will yield well before the concrete reaches its ultimate strength, the value of the nominal moment M_n can be written as

$$M_n = T\left(d - \frac{a}{2}\right) = A_s f_y \left(d - \frac{a}{2}\right) \qquad 10\text{-}4$$

and the usable flexural strength is

$$M_u = \phi M_n = \phi A_s f_y \left(d - \frac{a}{2}\right) \qquad 10\text{-}5$$

Substituting in this expression the value of a previously obtained gives an alternate form of M_u:

$$M_u = \phi A_s f_y d \left(1 - \frac{1}{1.7} \frac{\rho f_y}{f'_c}\right) \qquad 10\text{-}6$$

Maximum permissible steel percentage

If a balanced beam (neither underreinforced nor overreinforced) is used, it will theoretically fail suddenly and without warning. Accordingly, the ACI code limits the percentage of steel used in singly reinforced concrete beams (without axial loads) to 0.75 times the percentage that would produce balanced strain conditions.

In this section an expression is given for ρ_b, the percentage of steel required for a balanced design:

$$\rho_b = \left(\frac{0.85\beta_1 f'_c}{f_y}\right)\left(\frac{600}{600 + f_y}\right) \qquad 10\text{-}7$$

An overreinforced beam will fail in a brittle manner, while an underreinforced one will fail in a ductile manner. To attempt to make sure that only ductile failures can occur, the code, as previously mentioned, limits the maximum steel percentage as follows:

$$\rho_{max} = 0.75\rho_b \qquad 10\text{-}8$$

Mimimum percentage of steel

Sometimes because of architectural or functional requirements, beam dimensions are selected that are much larger than are required for bending alone. Such members theoretically require very small amounts of reinforcing.

There is actually another possible mode of failure that can occur in very lightly reinforced beams. If the ultimate resisting moment of the section is less than its cracking moment, the section will fail immediately when a crack occurs. This type of failure may occur without warning. To prevent such a possibility, the ACI code specifies a certain minimum amount of reinforcing that must be used at every section of flexural members where tensile reinforcing is required by analysis, whether for positive or negative moments. In the equations to follow, b_w represents the web width of beams.

$$A_{s,\min}=\frac{3\sqrt{f'_c}}{f_y}b_w d,\left(\text{in SI units } A_{s,\min}=\frac{\sqrt{f'_c}}{4f_y}b_w d\right) \tag{10-9}$$

$$\text{nor less than } \frac{200b_w d}{f_y}\left(\text{in SI units } \frac{1.4b_w d}{f_y}\right)$$

Expressing these values as percentages, $\rho_{\min}=\left(\frac{3\sqrt{f'_c}}{f_y}\right)$ but not less than $\frac{200}{f_y}$.

Or in SI units $\rho_{\min}=\frac{\sqrt{f'_c}}{4f_y}$ but not less than $\frac{1.4}{f_y}$.

Words and Expressions

derivation	[ˌderiˈveiʃən]	*n.* 推导,引出,来历
elongate	[ˈiːləŋgeit]	*v.* 拉长,(使)伸长 *adj.* 伸长的
fiber	[ˈfaibə]	*n.* 纤维
creep	[kriːp]	*n.* 蠕变,徐变
derive	[diˈraiv]	*v.* 推理出,推导,起源
rectangle	[ˈrekˌtæŋgl]	*n.* 长方形,矩形
parabola	[pəˈræbələ]	*n.*[数]抛物线
trapezoid	[ˈtræpizɔid]	*n.*[数]梯形,不等边四边形
centroid	[ˈsentrɔid]	*n.* 质心
equivalent	[iˈkwivələnt]	*adj.* 等效的,等价的,相当的 *n.* 等价物，相等物
couple	[ˈkʌpl]	*n.*[物]力偶,配偶
nominal	[ˈnɔminəl]	*adj.* 名义上的,[语]名词性的
substituting	[ˈsʌbstitjuːtiŋ]	*n.* 取代，代入，替代
balanced steel ratio		界限配筋率
underreinforced beam		适筋梁(即少于界限配筋的梁)
overreinforced beam		超筋梁
ultimate load		极限荷载
running time		逃生时间
in proportion to		与……成比例
stress diagram		应力图
equivalent rectangular stress block		等效矩形应力图形
SI units		国际(单位)制
resisting moment		抵抗力矩
flexural member		挠曲杆件

Notes

① underreinforced beam 为适筋梁,不是少筋梁;英语当中的少筋梁可以表示为 scarcely reinforced beam。

② Obviously, overreinforcing is a situation to be avoided if at all possible, and the code, by limiting the percentage of tensile steel that may be used in a beam, ensures the design of underreinforced beams and thus the ductile type of failures that provide adequate "running time".
显然,超筋是一种必须要避免的情况。在规范中,通过控制梁中受拉钢筋的配筋率,确保设计为适筋,使破坏为延性,为逃生提供足够的时间。

③ The actual stress variation and the actual location of the neutral axis vary somewhat from beam to beam depending on such variables as the magnitude and history of past loadings, shrinkage and creep of the concrete, size and spacing of tension cracks, speed of loading, and so on.
梁和梁之间实际应力和中和轴位置的不同,取决于加载历史、混凝土的收缩和徐变、受拉裂缝的尺寸和间距以及加载速度等的变化。

Exercises

Ⅰ. Translate the following words into Chinese.

1. balanced steel ratio ______________________________
2. underreinforced beams ______________________________
3. overreinforced beams ______________________________
4. ultimate load ______________________________
5. equivalent rectangular stress block ______________________________
6. strengths up to and including 30 MPa ______________________________
7. tensile reinforcing ______________________________

Ⅱ. Translate the following words into English.

1. 中性轴______________________________
2. 质心______________________________
3. 名义上的______________________________
4. 抵抗力矩______________________________
5. 最小配筋百分率______________________________
6. 最大配筋百分率______________________________

Ⅲ. Translate the following sentences into Chinese.

1. A beam that has a balanced steel ratio is one for which the tensile steel will theoretically start to yield and the compression concrete reach its ultimate strain at exactly the same load.
2. Tests of reinforced concrete beams confirm that strains vary in proportion to

distances from the neutral axis even on the tension sides and even near ultimate loads.

3. Based on these assumptions regarding the stress block, statics equations can easily be written for the sum of the horizontal forces and for the resisting moment produced by the internal couple.

Ⅳ. Translate the following sentences into English.

1. 在得到梁的表达形式之前,必须要根据两种受拉钢筋的配置数量对梁进行分类。
2. 对于一个适筋梁,当达到极限荷载时,梁的钢筋将开始屈服,虽然受压混凝土仍然没达到极限应力状态。
3. 有时,由于建筑和功能的要求,所选梁的尺寸比仅仅考虑弯曲要大。
4. 在应变达到某个值时,钢筋混凝土梁受拉边的混凝土出现开裂。
5. 当受拉边配筋较多时,受压边的破坏将以混凝土压碎为特征。

Ⅴ. Answer the following questions briefly.

1. Please tell the differences between underreinforced beams and overreinforced beams.
2. What is M_n? And what about M_u?
3. In which ways will beam fail after carrying capacity is reached? Tell the main differences in your own words.

Reading Materials

Behavior of Beams with Web Reinforcement

Due to inclined cracking, the strength of beams drops below the flexural capacity. The purpose of *web reinforcement* (腹筋) is to ensure that the full flexural capacity can be developed.

Prior to inclined cracking, the strain in the *stirrups* (箍筋) is equal to the corresponding strain of the concrete. Since concrete cracks at a very small strain, the stress in the stirrups prior to inclined cracking will not exceed 3 to 6 ksi. Thus stirrups do not prevent inclined cracks from forming; they come into play only after the cracks have formed.

The forces in a beam with stirrups and all inclined crack are shown in Fig. 10-4. The shear transferred by tension in the stirrups is V_s. Since V_s does not disappear when the crack opens, there will always be a compression force C_1' and a shear force V_{cz}' acting on the part of the beam below the crack. As a result, T_2 will be less than T_1, the difference depending on the amount of web reinforcement. The force T_2 will, however, be larger than $T=M/jd$ based on the moment at C.

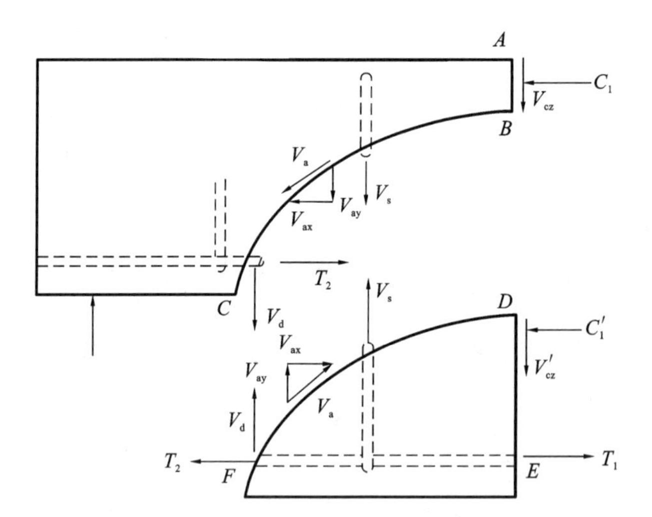

Fig. 10-4 Internal force in a cracked beam with stirrups

The loading history of such a beam is shown qualitatively in Fig. 10-5. Prior to flexural cracking, all the shear is carried by the uncracked concrete. Between flexural and inclined cracking, the external shear is resisted by V_{cz}, V_{ay}, and V_d. Eventually, the stirrups crossing the crack yield, and V_s stays constant for higher applied shears. Once the stirrups yield, the inclined crack opens more rapidly. As the inclined crack widens, V_{ay} decreases further, forcing V_d and V_{cz} to increase at an accelerated rate until either a *splitting*(爆裂似的，极快的)(dowel) failure occurs, or the compression zone crushes due to combined shear and compression.

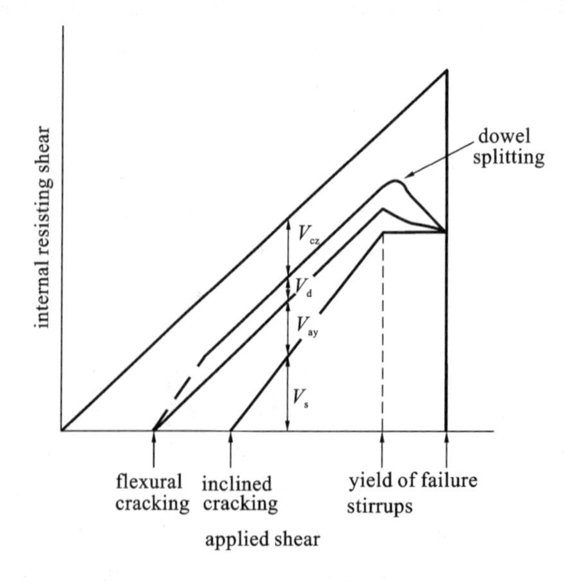

Fig. 10-5 Distribution of internal shears in a beam with web reinforcement

Each of the components of this process except V_s has a brittle load-deflection response. As a result, it is difficult to quantify the contributions of V_{cz}, V_d, and V_{ay}.

In design these are *lumped*(汇集) together as V_c, referred to somewhat incorrectly as "the shear carried by the concrete". Thus the nominal shear strength, V_n, is assumed to be

$$V_n = V_c + V_s \tag{10-10}$$

Traditionally in North American design practice, V_c is taken equal to the failure capacity of a beam without stirrups, which, in turn, is taken equal to the inclined cracking shear as suggested by the line indicating inclined cracking and failure for a/d from 2.5 to 6.5.

The behavior of beams failing in shear must be expressed in terms of a mechanical-mathematical model before designers can make use of this knowledge in design. The best model for beams with web reinforcement is the truss model. This is applied to slender beams and to *deep beams*(深梁).

In 1899 and 1902, respectively, the Swiss engineer Ritter and the German engineer Mörsch, independently, published papers proposing the truss analogy for the design of reinforced concrete beams for shear. These procedures provide an excellent *conceptual*(概念上的) model to show the forces that exist in a cracked concrete beam.

As shown in Fig. 10-6(a), a beam with inclined cracks develops compressive and tensile forces, C and T, in its top and bottom "*flanges*(翼缘)", vertical tensions in the stirrups and inclined compressive forces in the concrete "diagonals" between the inclined cracks. This highly indeterminate system of forces is replaced by analogous truss. The simplest truss is shown in Fig. 10-6(b).

Several assumptions and simplifications are needed to derive the analogous truss. In Fig. 10-6(b) the truss is formed by lumping all of the stirrups cut by section A-A into one vertical member b-c and all the diagonal concrete members cut by section B-B into one diagonal member e-f. This diagonal member is stressed in compression to resist the shear on section B-B. The compression chord along the top of the truss is actually a force in the concrete but is shown as a truss member. The compressive members in the truss are shown with dashed lines to imply that they are really forces in the concrete, not separate truss members. The tensile members are shown with solid lines.

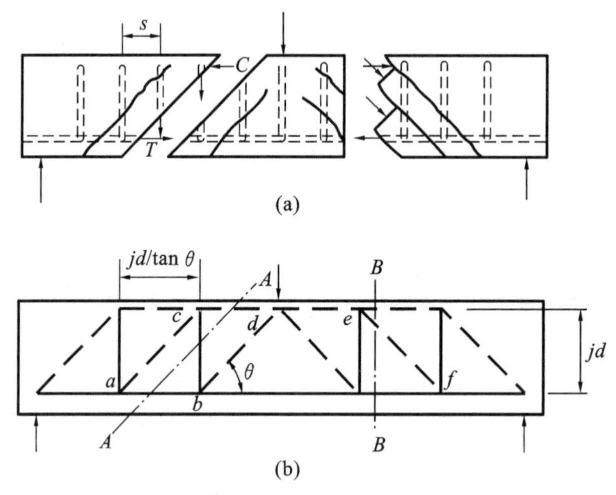

Fig. 10-6 Truss analogy

(a)internal forces in a cracked beam; (b)pin-jointed truss

土木工程专业英语论文写作技巧(1)——论文中的“引用”

在进行学术写作时,常常需要引用他人的文献,即引用。英语将引用分为“引用”(citation)和“引语”(quotation)。其中“引用”另一位作者的意见,即意味着你用自己重新组织的语言告诉读者他人的学术论点。而“引语”则是指使用被引作者的原话,并在原话前后加引号。同时请注意,若在“引语”中使用冒号,则在引语前不需要单词“that”。引语(quotation)的使用实例如下。

Sachs concludes that “The idea of development stands today like a ruin in the intellectual landscape…” (Sachs, 1992a, p. 156)(注意:这里使用了 that,后面不用冒号。)

In his paper, J. Bechtold explains: “In the case of earthquakes the danger accrues from the rigid foundation of the buildings…”(注意:这里使用了冒号,前面不用“that”。)

上述“explains,concludes”也可以根据论文含义和时态换成“writes”“maintains”和“argued”等。

实际上,在科技文章中引用(citation)比引语(quotation)更常用,作者往往将他人的学术观点和成果融合到自己的段落和上下文中,这样行文更流畅,逻辑也更严谨。例如在“Moreover, the floor slab may not be as continuous as in an in-situ build, and load transfer between adjacent modules may occur mainly through the module corners[11].”中,作者已经将他人的学术观点(该文章中参考文献 11 的观点)用自己的语言重写,并与上下文组成有机的整体。

Lesson 11

INTRODUCTION OF BEAMS AND COLUMNS

Beams

Beams are among the most common members that one will find in structures. They are structural members which carry loads that are applied at right angles to the longitudinal axis of the member. This causes the beam to bend. In this article we consider beams that carry no axial force. Fig. 11-1(a) and (b) illustrate some typical examples of beam applications.

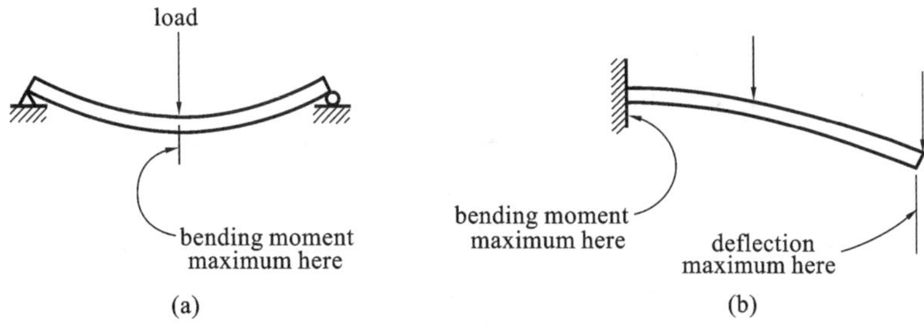

Fig. 11-1 Beam types

(a)simply supported beam; (b)cantilever beam

When visualizing a beam (or any structural member) for the purposes of analysis or design, it is convenient to think of the member in some idealized form. The idealized form represents as closely as possible the actual structural member, but it has the advantage that it can be dealt with mathematically. For instance, in Fig. 11-1(a), the beam is shown with simple supports, a pin (knife-edge or hinge) on the left and a roller on the right, create conditions which are easily treated mathematically when it becomes necessary to find beam reactions, shears, moments, and deflections. Recall that the pin support will provide vertical and horizontal reactions (but no resistance to rotation) and the roller will provide only a vertical reaction. This is particularly significant for bridges where provisions must be made for expansion and contraction due to temperature changes. In buildings, generally each support is capable of furnishing vertical and horizontal reactions. However, the beams are still considered to be simply supported since the requirement of a simple support on the left side. This type of support provides vertical and horizontal reactions, as well as resistance (or a reaction) to rotation. The one fixed support is sufficient for static equilibrium of the beam. Although the idealized conditions generally will not exist in the actual structure, the actual conditions will approximate the ideal conditions and should be close enough to allow for a reasonable analysis or design.[①]

In the process of beam design, we will be concerned initially with the bending moment in the beam. The bending moment is produced in the beam by the loads it supports. Other effects, such as shear or deflection, may eventually control the design of the beam and will have to be checked. But usually moment is critical and it is, therefore of initial concern.

Beams are sometimes called by other names, which are indicative of some specialized functions.

girder: a major, or deep beam which often provides support for other beams.

stringer: a main longitudinal beam, usually in bridge floors.

floor beam: a transverse beam in bridge floors.

joist: a light beam that supports a floor.

lintel: a beam spanning across an opening (a door or a window), usually in masonry construction.

spandrel: a beam on the outside perimeter of a building which supports, among other loads, the exterior wall.

purlin: a beam that supports a roof, and frames between or over supports, such as roof trusses or rigid frames.

girt: generally, a light beam that supports only the lightweight exterior sides of a building (typical in pre-engineered metal buildings).

Columns

Structural members that carry compressive loads are sometimes given names which identify them as to their function. Compression members that serve as bracing are commonly called struts. Other compression members may be called posts and pillars. Trusses are composed of members that are in compression and members that are in tension. These may be either chord or web members. Of primary interest in this article will be the main vertical compression members in building frames which are called columns. Additionally, double angle compression members will be discussed. Columns are compression members which have their length dimension considerably larger than their least cross-sectional dimension.

In this article we consider members are subjected to axial (concentric) loads; that is the loads are coincident with the longitudinal centroidal axis of the member. This is a special case and one that exist rarely, if at all. However, where small eccentricities exist, it may be assumed that an appropriate factor of safety will compensate for the eccentricity and the column may be designed as though it were axially loaded.[②] If we consider the range of possible combination of load and moment supported on columns, then at one end of the range is the axially loaded column.

This column carries no moment. At the other end of the range is the member that carries only moment with no (or very little) axial load. (As a moment-carrying member, it could be considered to be a beam.) When a column carries both axial load and moment, it is called beam-column.

Commonly used cross sections for steel compression members include most of the rolled shapes. For larger loads it is common to use a built-up cross section. In addition to providing increased cross sectional area, the built-up sections allow a designer to tailor to specific needs the radius of gyration(r) values about the x-x and y-y axes. ③ Typical compression cross sections are shown in Fig. 11-2. The dashed lines shown on the cross sections of Fig. 11-2(f) and (g) represent tie plates, lacing bars, or perforated cover plates and do not contribute to the cross sectional properties. Their function is to hold the main longitudinal components of the cross section in proper relative position and to make the built-up section act as a single unit.

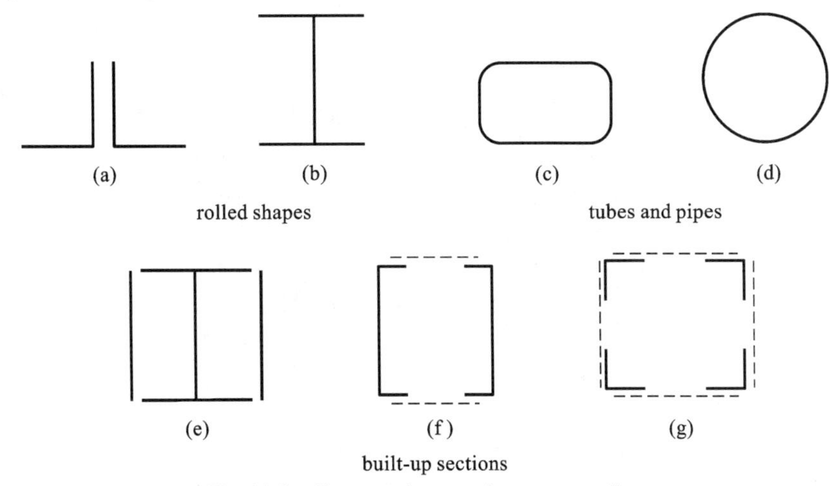

Fig. 11-2 Compression member cross sections

In dealing with compression members, the problem of stability is of great importance. Unlike tension members, where the load tends to hold the members in alignment, compression members are very sensitive to factors that may tend to cause lateral displacement or buckling. ④ The situation is similar in ways to the lateral buckling of beams. The buckling problem is intensified and the load-carrying capacity is affected by such factors as eccentric load, imperfection of material, and initial crookedness of the member. Also, residual stresses play a role. These are the variable stresses that are "locked up" in the member as a result of the method of manufacture, which involves unequal cooling rates within the cross section. The parts that cool first (such as the flange tips) will have residual compression stresses. While parts that cool last (the junction of the flange and web) will have residual tension stresses. Residual stresses may also be induced by nonuniform plastic deformation caused by cold working, such as in the straightening process and the cambering process.

Words and Expressions

pin	[pin]	*n*. 销钉(栓)
hinge	[hindʒ]	*n*. 铰链
		v. 铰接
roller	['rəulə]	*n*. 滚轴
provision	[prəu'viʒən]	*n*. 预防,条款
contraction	[kən'trækʃən]	*n*. 收缩,压缩
furnish	['fə:niʃ]	*v*. 提供
transverse	[trænz'və:s]	*n*. 横断面
		adj. 横截面的,横向的
joist	[dʒɔist]	*n*. 搁栅,托梁
lintel	['lintl]	*n*. 楣,过梁;(连接墙肢的)连梁
purlin	['pə:lin]	*n*. 檩条
truss	[trʌs]	*n*. 桁架
girt	[gə:t]	*n*. 围梁,箍梁
chord	[kɔ:d]	*n*. 弦,弦杆
concentric	[kən'sentrik]	*adj*. 同心的,同轴的
centroidal	[sen'trɔidəl]	*adj*. 形心的
eccentricity	[ˌeksen'trisəti]	*n*. 偏心距
gyration	[ˌdʒaiə'reiʃən]	*n*. 旋转,回转
perforate	['pə:fəreit]	*v*. 穿(钻)孔,打眼
stability	[stə'biləti]	*n*. 稳定(性)
imperfection	[ˌimpə'fekʃən]	*n*. 缺陷
crooked	['krukid]	*adj*. 歪的,扭曲的
flange	[flændʒ]	*n*. 翼缘
junction	['dʒʌŋkʃən]	*n*. 连接,接合处
straighten	['streitən]	*v*. 矫直,整平
camber	['kæmbə]	*v*. 翘曲,弯曲,拱起
longitudinal axis		纵轴
masonry construction		砌体结构
beam-column		压弯构件
built-up section		组合截面
tie plate		缀板
lacing bar		缀条
in alignment		成一直线
residual stress		残余应力

Notes

① Although the idealized conditions generally will not exist in the actual structure, the actual conditions will approximate the ideal conditions and should be close enough to allow for a reasonable analysis or design.
虽然在实际结构中理想化的状态一般不存在，但实际的条件可以近似并且应该足够地接近理想状态以供合理的分析和设计之用。

② However, where small eccentricities exist, it may be assumed that an appropriate factor of safety will compensate for the eccentricity and the column may be designed as though it were axially loaded.
然而，当存在小偏心时，用适当的安全系数可以补偿偏心，这样柱可以按轴心受力来设计。

③ In addition to providing increased cross sectional area, the built-up sections allow a designer to tailor to specific needs the radius of gyration(r) values about the x-x and y-y axes.
除了可以增大横截面积之外，设计师还可以采用组合截面以满足(截面)关于x-x轴和y-y轴的回转半径的需求。

④ Unlike tension members, where the load tends to hold the members in alignment, compression members are very sensitive to factors that may tend to cause lateral displacement or buckling.
与受拉构件中荷载使构件保持在同一直线上不同，受压构件对于能引起侧向位移和屈曲的因素很敏感。

Exercises

Ⅰ. Translate the following words into Chinese.

1. vertical and horizontal reactions __________
2. longitudinal centroidal axis __________
3. lateral displacement or buckling __________
4. nonuniform plastic deformation __________
5. centroidal axis __________

Ⅱ. Translate the following words into English.

1. 温度变化引起的膨胀和收缩__________
2. 静力平衡__________
3. 冷作、冷加工__________
4. 安全系数__________
5. 荷载组合__________
6. 回转半径__________
7. 压弯构件__________

Ⅲ. Translate the following sentences into Chinese.

1. The idealized form represents as closely as possible the actual structural member, but it has the advantage that it can be dealt with mathematically.
2. Other effects, such as shear or deflection, may eventually control the design of the beam and will have to be checked. But usually moment is critical and it is, therefore of initial concern.
3. Columns are compression members which have their length dimension considerably larger than their least cross-sectional dimension.
4. Their function is to hold the main longitudinal components of the cross section in proper relative position and to make the built-up section act as a single unit.
5. These are the variable stresses that are "locked up" in the member as a result of the method of manufacture, which involves unequal cooling rates within the cross section.

Ⅳ. Translate the following sentences into English.

1. 钢结构有三种构件连接方式:焊接、铆接和螺栓连接。
2. 钢箱梁具有较大的截面尺寸及刚度,是一种较理想的桥梁结构构件。
3. 钢结构构件的稳定性是设计的一个重要方面,在工程实际中应给予足够的重视。
4. 在钢结构中,柱构件常常是压弯构件,即同时承受轴力和弯矩的作用。

Ⅴ. Answer the following questions briefly.

1. What is the definition of beam?
2. How should we consider bending moment and other effects in the process of beam design?

Reading Materials

Tied and Spiral Columns

Over 95% of all columns in buildings in nonseismic regions are tied columns. *Tied columns*(箍筋柱) may be square, rectangular, L shaped, circular, or any other required shape. Occasionally when high strength and/or high ductility are required, the bars are placed in a circle and the ties are replaced by a bar bent into a helix or spiral, with a pitch of $1\frac{3}{8}$ to $3\frac{3}{8}$ in.

Such a column, called a *spiral column*(螺旋箍筋柱). Spiral columns are generally circular, although square or *polygonal*(多角形的) shapes are sometimes used. The spiral acts to restrain the lateral expansion of the column *core*(核心) under axial loads causing crush, and in doing so, delays the failure of the core, making the column more ductile. In seismic regions, the ties are heavier and much more closely spaced. Spiral columns are used more extensively in such regions.

Fig. 11-3(a) shows a portion of the core of a spiral column enclosed by one and a half turns of the spiral. Under compressive loads the concrete in this column shortens longitudinally under the stress f_1, and due to Poisson's ratio it expands laterally. This lateral expansion is especially pronounced at stresses in excess of 70% of the *cylinder strength*(圆柱体强度). In a spiral column, the lateral expansion of the concrete inside the spiral (referred to as the core) is restrained by the spiral. This stresses the spiral in tension as shown in Fig. 11-3(b). For equilibrium the concrete is subjected to lateral compressive stresses f_2. An element taken out of the core (see Fig. 11-3(c)) is subjected to triaxial compression; triaxial compression is shown to increase the strength of concrete:

$$f_1 = f_c + 4.1 f_2 \tag{11-1}$$

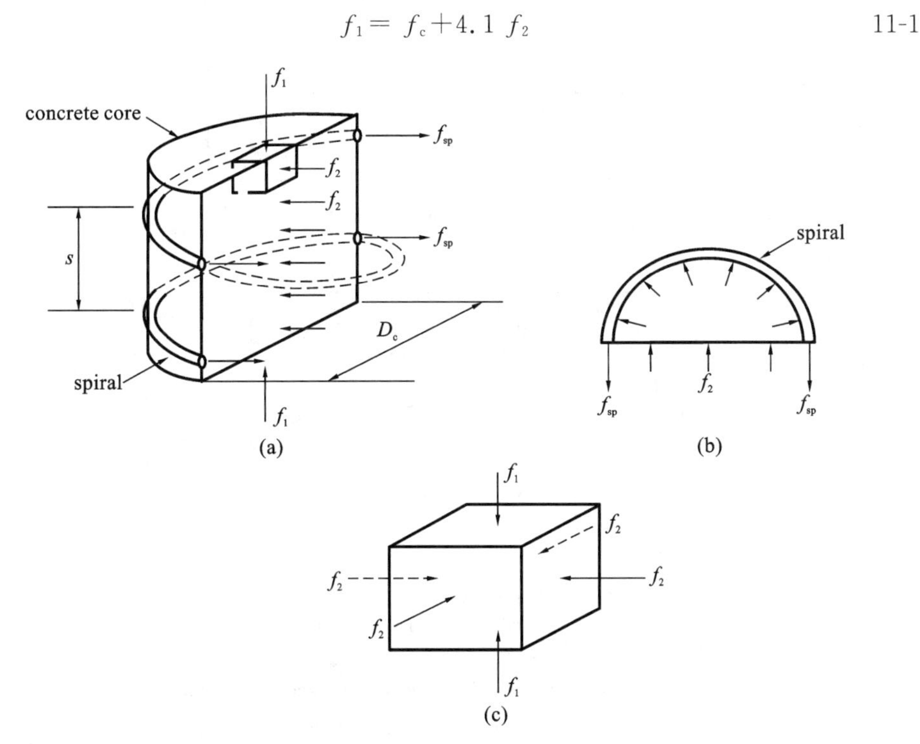

Fig. 11-3 Triaxial stresses in core of spiral column

In a tied column in a nonseismic region, the ties are spaced roughly the width of the column apart, and as a result, provide relatively little lateral restrain to the core. Outward pressure on the sides of the ties due to lateral expansion of the core merely bends them outward, developing a *negligible*(可以忽略的) *hoop stress*(环向应力)effect. Hence normal ties have little effect on the strength of the core in a tied column. They do, however, act to reduce the *unsupported length*(自由长度) of the longitudinal bars, thus reducing the danger of buckling of those bars as the bar stress approaches yield.

Fig. 11-4 presents load deflection diagrams for a tied column and a spiral column subjected to axial loads. The initial parts of these diagrams are similar. As the maximum load is reached, vertical cracks and crushing develop in the concrete shell outside the ties or spiral, and this concrete *spalls off*(剥落). When this occurs in a tied column, the capacity of the core that remains is less than the load and the concrete core crushes and the reinforcement buckles outward between ties. This occurs suddenly, without warning, in a brittle manner.

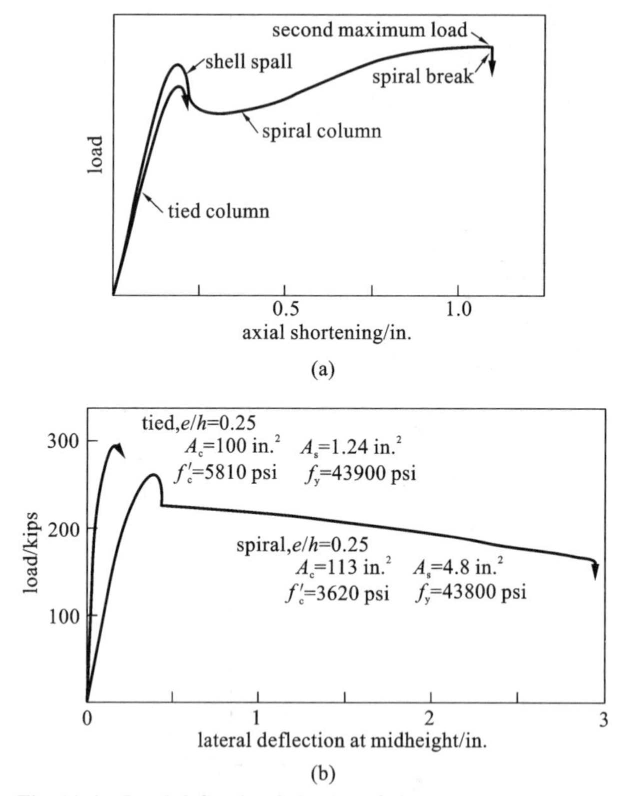

Fig. 11-4 Load-deflection behavior of tied and spiral columns

(a) axially loaded columns; (b) eccentrically loaded columns

When the shell spalls off the spiral column, the column does not fall immediately because the strength of the core has been enhanced by the *triaxial*(三轴的) stresses resulting from the effect of the spiral reinforcement. As a result, the column can undergo large deformations, eventually reaching a second maximum load when the spirals yield and the column finally collapses. Such a failure is much more ductile and gives warning of the impending failure together with possible load redistribution to other members. It should be noted, however, that this is accomplished only at very high strains. For example, the strains necessary to reach the second maximum load correspond to a shortening of about 1 in. in an 8-ft-high column, as shown in Fig. 11-4(a). When spiral columns are eccentrically loaded, the second maximum load may be less than the initial maximum, but the deformations at failure are still large, allowing load redistribution (see Fig. 11-4(b)). Because of their greater ductility, spiral columns are assigned a capacity reduction factor, ϕ, of 0.75, rather than the value of 0.70 used for tied columns.

Spiral columns are used when ductility is important or where high loads make it economical to utilize the extra strength resulting from the higher ϕ factor. Fig. 11-5 and Fig. 11-6 show a tied and a spiral column, respectively, after an earthquake. Both columns are in the same building and have undergone the same deformations. The tied column has failed completely, while the spiral column, although badly damaged, is still supporting a load. The very minimal ties in Fig. 11-5 were inadequate to confine the core concrete. Had the column contained ties detailed according to ACI Sec. A4.4, it would have performed much better.

Fig. 11-5 Tied column destroyed in 1971 San Fernando earthquake

Fig. 11-6 Spiral column damaged by 1971 San Fernando earthquake

土木工程专业英语论文写作技巧(2)——“负迁移”

学习英语不仅是记单词,并将这些单词融入母语语法中,还需要学习全新的表达方式。我们最好摒弃在母语中已经形成的习惯。母语的习惯会对我们学习新语言产生一定的影响,这种影响称为语言迁移。

对于母语为汉语的人来说,学会构建简单的英语句子比较容易,因为汉语和英语有相同的基本语序,即主谓宾。例如,我们可以根据汉语句子“我吃苹果”,逐词翻译,从而得到一个语法正确的英语句子。这种情况下母语的语法知识帮助我们形成一个正确英语表达,这种情况称为正迁移。相反地,当母语的知识导致我们产生语法不正确的英语表达时,称为“负迁移”(negative transfer)。

例如,下面两种常用的汉语句子结构,应该避免在英语中使用。

· 因为……所以……(英语中常见的错误为:because…, so…)

· 虽然……但是……(英语中常见的错误为:although…,but…)

because…, so…或者 although…, but…像汉语那样成对使用并不符合英语语法,他们的每一对只能使用其中一个。例如,because 和 so 不应该一起使用,因为它们具有相同的功能:在论述原因和结果时使用。because 在原因之前使用,so 在结果之前使用。例如下面两句为正确的表达(分别单独用 because 和 so 表示原因和结果)。

Difficulty in modeling connection behavior arises *because* almost all connections behave inelastically and therefore nonlinearly.

· Gravity load is resisted by both the internal columns and peripherial diagonal columns, *so* internal columns need to be designed for vertical load only.

而对于 although 和 but,不要在同一句中同时用这两个词。请注意,although 总是在第一个子句之前,而 but 总是介于两个子句之间(不在句子的开头使用 but)。请阅读下列两句。

· *Although* the method produces accurate fits, its use in computer programs would require the storing of large amounts of data.

· By taking a sufficient number of terms, almost any M-9 curve can be closely fitted, *but* the parameters have very little physical meaning.

because 和 although 是从属连词,在英语句子里使用这两个词,会使一个子句依赖于另一个子句。母语为英语的读者在看到这两个词的时候习惯去找另外一个子句,两个子句共同构成一个完整的表达。而我们有汉语习惯的读者见到这两个连词的时候习惯去找搭配的另一半连词。

so 和 but 是协调连词,它们被用来连接同级别的元素,连接单词到单词、短语到短语、从句到从句。例如:

· I am curious about the Golden Age of Mexican cinema, *so* I listen to Pedro Infante's music.

· I listen to Bollywood music, *but* I do not understand Hindi.

综上所述,语言迁移是学习英语的一部分,而负迁移可能影响任何曾经学习过一门新语言的人,我们一定要重视这种现象并尽量避免。常用的负迁移有如下表现:①主谓不一致;②名词单复数用错;③冠词误用;④词性误用等。

Lesson 12

EFFECTIVE LENGTHS OF COLUMNS

Long, short, and intermediate columns

A column subject to an axial compression load will shorten in the direction of the load. If the load is increased until the column buckles, the shortening will stop, and the column will suddenly bend or deform laterally and may at the same time twist in a direction perpendicular to its longitudinal axis. ①

The strength of a column and the manner in which it fails are greatly dependent on its length. A very short stocky steel column may be loaded until the steel yields and perhaps on into the strain-hardening range. As a result, it can support about the same load in compression that it can in tension.

As the length of a column increases, its buckling stress will decrease. If the length exceeds a certain value, the buckling stress will be less than that of the steel. Columns in this range are said to fail elastically.

Long steel columns will fail at loads which are proportional to the bending rigidity of the column (EI) and independent of the strength of the steel. For instance, a long column constructed with a 36 ksi yield stress steel will fail at just about the same load as one constructed with a 100 ksi yield stress steel.

Columns are sometimes classed as being long, short, or intermediate. A brief discussion of each of these classifications is presented in the paragraphs to follow.

(1) Long columns.

The Euler formula predicts very well the strength of long columns where the axial buckling stress remains below the proportional limit. Such columns will buckle elastically.

(2) Short columns.

For very short columns the failure stress will equal the yield stress and no buckling will occur. (For a column to fall into this class it would have to be so short as to have no practical application. Thus no further reference is made to them here.)

(3) Intermediate columns.

For intermediate columns some of the fibers will reach the yield stress and some will not. The members will fail by both yielding and buckling, and their behavior is said to be inelastic. (For the Euler formula to be applicable for such columns it would have to be modified according to the reduced modulus concept or the tangent modulus concept to account for the presence of residual stresses.[②]) The vast majority of columns fall into this class.

End restraint and effective lengths of columns

End restraint and its effect on the load-carrying capacity of columns is a very important subject indeed. Columns with appreciable end restraint can support considerably more load than those with little end restraint, as at hinged ends.

The effective length of a column is the distance between points of zero moment in the column, that is, the distance between its inflection points. In steel specifications the effective length of a column is referred to as KL, where K is the effective length factor. K is the number that must be multiplied by the length of the column to obtain its effective length. Its size depends on the rotational restraint supplied at the ends of the column and upon the resistance to lateral movement provided.

The concept of effective lengths is simply a mathematical method of taking a column whatever its end and bracing conditions and replacing it with an equivalent pinned-end braced column. A complex buckling analysis could be made for a frame to determine the critical stress in a particular column. The K factor is determined by finding the pinned-end column with an equivalent length that provides the same critical stress. The K factor procedure is a method of making simple solutions for complicated frame buckling problems.

Columns with different end conditions have entirely different effective lengths. For this initial discussion it is assumed that no sidesway or joint translation is possible (Sidesway or joint translation means that one or both ends of a column can move laterally with respect to each other). Should a column be connected with frictionless hinges as shown in Fig. 12-1(a), its effective length would be equal to the actual length of the column, and K would equal 1.0. If there were such a thing as a perfectly fixed-end column, its points of inflection (or points of zero moment) would occur at its one-fourth points, and its effective length would equal $L/2$ as shown in Fig. 12-1(b). As a result, its K value would equal 0.50.[③]

Obviously the smaller the effective length of a particular column, the smaller its danger of lateral buckling and the greater its load-carrying capacity. In Fig. 12-1 (c) a column is shown with one end fixed and one end pinned. The K for this column is theoretically 0.70.

Actually there are no perfect pin connections or any perfect fixed ends, and the usual column in a braced frame falls between the two extremes. This discussion would seem to indicate that column effective lengths always vary from an absolute minimum of $L/2$ to an absolute maximum of L, but there are exceptions to this rule. An example is given in Fig. 12-2(a) where a simple bent is shown. The base of each of the columns is pinned, and the other end is free to rotate and move later-

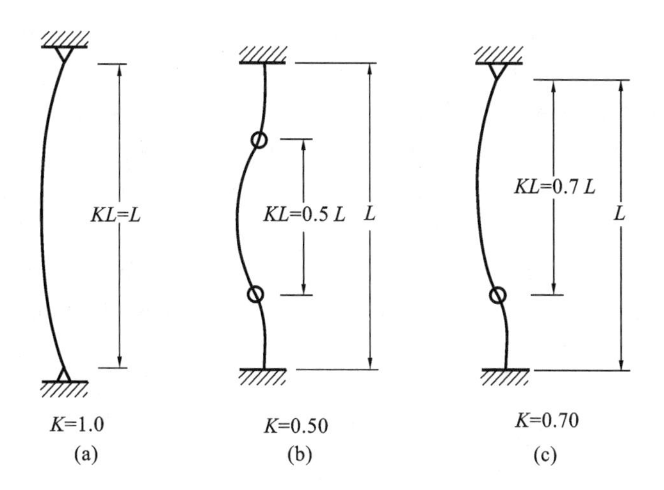

Fig. 12-1 Effective lengths for columns in braced frames (sidesway prevented)

(a) pinned end column; (b) fixed end column; (c) one end fixed and one end pinned column

ally (sidesway). Examination of this figure will show that the effective length will exceed the actual length of the column, as the elastic curve will theoretically take the shape of the curve of a pinned-end column of twice its length and K will theoretically equal 2. 0. ④ Notice in part (b) of the figure how much smaller the lateral deflection of column AB would be if it were pinned both top and bottom so as to prevent sidesway.

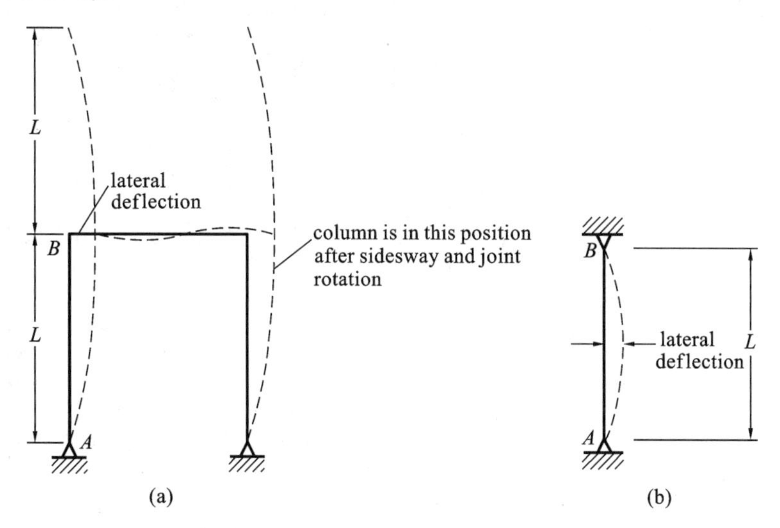

Fig. 12-2 Effective length of a sway frame

(a) lateral deflection shape of a sway frame; (b) lateral deflection shape of a pinned end column

Structural steel columns serve as parts of frames, and these frames are sometimes braced and sometimes unbraced. In a braced frame sidesway or joint translation is prevented by means of bracing, shear walls, or lateral support from adjoining structures. An unbraced frame does not have any of these types of bracing and must depend on the bending stiffness of beams and columns rigidly connected to each other to prevent lateral buckling. For braced frames K values can never be greater than 1.0, but for unbraced frames the K values will always be greater than 1.0 because of sidesway.

Table 12-1 of the Commentary on the ASD Specification gives recommended effective length factors when ideal conditions are approximated. This table is reproduced here as Table 12-1. Two sets of K values are provided in the table, one being the theoretical values and the other being the recommended design values based on the fact that perfectly pinned and fixed conditions are not possible. If the ends of the column of Fig. 12-1(b) were not quite fixed, the column would be a little freer to bend laterally and its points of inflection would be farther apart. The recommended design K given is 0.65 while the theoretical value is 0.5. As no column ends are perfectly fixed or perfectly hinged, the designer may wish to interpolate between the values given in the table, the interpolation to be based on his or her judgment of the actual restraint conditions.

Table 12-1 Column effective lengths

	(a)	(b)	(c)	(d)	(e)	(f)
Buckled shape of column shown by dashed line						
Theoretical K value	0.5	0.7	1.0	1.0	2.0	2.0
Recommended design value when ideal conditions are approximated	0.65	0.80	1.2	1.0	2.10	2.0
End condition code		rotation fixed and translation fixed				
		rotation free and translation fixed				
		rotation fixed and translation free				
		rotation free and translation free				

Words and Expressions

bend（过去式和过去分词 bent）	[bend]	*v.* 使弯曲，专心于，使屈服 *n.* 弯曲
twist	[twist]	*v.* 扭弯，扭曲 *n.* 扭曲，歪曲
perpendicular	[ˌpəːpənˈdikjulə]	*adj.* 垂直的
stocky	[ˈstɔki]	*adj.* 矮壮的，低矮结实的
factor	[ˈfæktə]	*n.* 系数
brace	[breis]	*n.* 支撑，支柱 *v.* 支撑
friction	[ˈfrikʃən]	*n.* 摩擦，摩擦力
approximate	[əˈprɔksimeit]	*adj.* 近似的，大约的 *v.* 近似，约计
interpolate	[inˈtəːpəleit]	*v.* 插值
effective length		计算长度
intermediate column		中长柱
strain-hardening range		应变硬化阶段，强化阶段
buckling stress		屈曲应力
bending rigidity(benaing stiffness)		抗弯刚度
long column		长柱
Euler formula		欧拉公式
proportional limit		比例极限
short column		短柱
reduced modulus		折算模量
tangent modulus		切线模量
end restraint		端部约束
load-carrying capacity		承载能力
hinged ends		铰支端
inflection point		反弯点，拐点，转折点
effective length factor		计算长度系数
critical stress		临界应力
sidesway		侧移
translation		平移，平动
braced frame (unbraced frame)		支撑框架(无支撑框架)
elastic curve		弹性曲线
sway frame		有侧移框架
lateral support		侧向支撑

Notes

① If the load is increased until the column buckles, the shortening will stop, and the column will suddenly bend or deform laterally and may at the same time twist in a direction perpendicular to its longitudinal axis.
如果外力一直增长直到柱发生屈曲,压缩将停止,柱子将突然发生弯曲或侧向变形,同时也可能在绕柱子轴线的方向发生扭转。

② For the Euler formula to be applicable for such columns it would have to be modified according to the reduced modulus concept or the tangent modulus concept to account for the presence of residual stresses.
由于残余应力的存在,为了使欧拉公式适用于这些梁,必须按照折减模量概念或切线模量概念来进行修正。

③ If there were such a thing as a perfectly fixed-end column, its points of inflection (or points of zero moment) would occur at its one-fourth points, and its effective length would equal $L/2$ as shown in Fig. 12-1(b). As a result, its K value would equal 0.50.
对于一个理想的固端支撑柱,其反弯点(或弯矩为零的点)在柱长的1/4处,它的计算长度为$L/2$(如图12-1(b)所示),因此K值等于0.50。

④ Examination of this figure will show that the effective length will exceed the actual length of the column, as the elastic curve will theoretically take the shape of the curve of a pinned-end column of twice its length and K will theoretically equal 2.0.
通过对数据的观察可以发现,由于理论上弹性弯曲能达到相当于其长度两倍的两端铰支柱的弯曲形状,故有效长度可以超过柱子的实际长度,此时K在理论上应等于2.0。

Exercises

Ⅰ. Translate the following words into Chinese.

1. axial compression load ______
2. a short stocky steel column ______
3. a column subject to an axial compression load ______
4. strain-hardening range ______
5. fail by both yielding and buckling ______
6. the reduced modulus concept ______
7. the tangent modulus concept ______

Ⅱ. Translate the following words into English.

1. 屈曲应力______
2. 比例极限______

3. 抗弯刚度____________________
4. 失稳分析____________________
5. 计算长度系数____________________
6. 铰支端____________________
7. 反弯点____________________
8. 节点平移____________________

Ⅲ. Translate the following sentences into Chinese.

1. A very short stocky steel column may be loaded until the steel yields and perhaps on into the strain-hardening range.
2. Long steel columns will fail at loads which are proportional to the bending rigidity of the column (*EI*) and independent of the strength of the steel.
3. The effective length of a column is the distance between points of zero moment in the column, that is, the distance between its inflection points.
4. Obviously the smaller the effective length of a particular column, the smaller its danger of lateral buckling and the greater its load-carrying capacity.
5. In a braced frame sidesway or joint translation is prevented by means of bracing, shear walls, or lateral support from adjoining structures.
6. As no column ends are perfectly fixed or perfectly hinged, the designer may wish to interpolate between the values given in the table, the interpolation to be based on his or her judgment of the actual restraint conditions.

Ⅳ. Translate the following sentences into English.

1. 柱子在轴向压力作用下沿力的方向变短。
2. 柱子发生强度破坏还是失稳破坏由柱子的长度所决定。
3. 事实上没有理想的铰接和固接,一般的支撑框架都在这两个极限之间屈曲。

Ⅴ. Answer the following questions briefly.

1. Please describe the characteristics of the three kinds of columns discussed in this article.
2. What is the effective length of a column?
3. What are the factors concerning the effective length of a column?

Reading Materials

Steel Connections

There are many different types of connections between structural steel members, depending on the type and structural use of the connection, as well as on the connectors, plates and other elements used in the connection. The methods of ordinary connection and their structural behavior are discussed in the following sub-sections.

Ordinary bolts

The "black" *hexagon head bolt*(六角头螺栓)shown in Fig. 12-3 with *nut*(螺母)and *washer*(垫圈)is the most commonly used structural fastener. The bolts are in two strength grades as specified in *BS*(英国规范)4190.

Grade 4.6: mild steel, yield stress=235 N/mm^2

Grade 8.8: high-strength steel, yield stress=627 N/mm^2

The main diameters used are(The sizes shown in brackets are not preferred).

16, 20, (22), 24, (27) and 30 mm

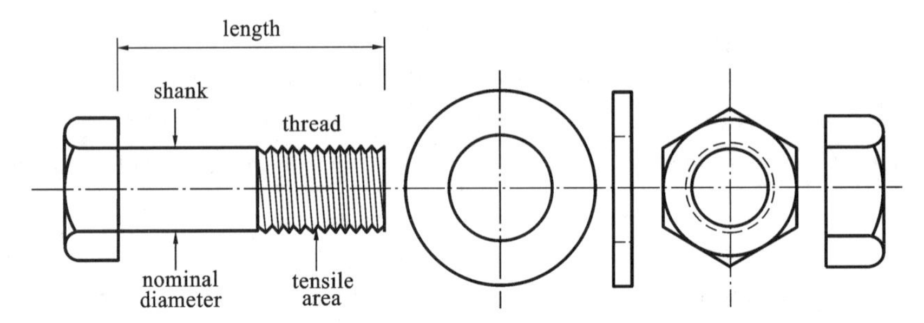

Fig. 12-3 Hexagon head bolt, nut and washer

Bolts may be arranged to act in single or double shear, as shown in Fig. 12-4. Provisions governing spacing, edge and end distances are set out in relevant standard. The principal provisions in normal conditions are as below.

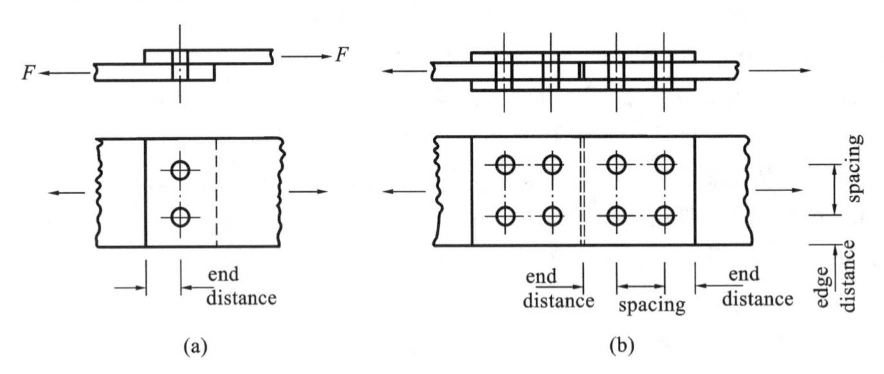

Fig. 12-4 Bolts in single and double shear

①The minimum spacing is 2.5 times the bolt diameter.

②The maximum spacing in unstiffened plates in the direction of stress is $14t$, where t is the thickness of the thinner plate connected.

③The minimum edge and end distance as shown in Fig. 12-4 from a *rolled*(轧制的), *machine-flame cut*(火焰切割的) or *planed*(刨边的)edge is $1.25D$, where D is the hole diameter.

④The maximum edge distance is $11t\varepsilon$, where $\varepsilon=(275/p_y)^{0.5}$.

The clause should be consulted for requirements in corrosive areas.

Note that the nominal diameters of holes for ordinary bolts are greater than the bolt diameter by: 2 mm for bolts up to 22 mm diameter and 3 mm for larger-diameter bolts.

A shear joint can fail in the following four ways:

①By shear on the bolt shank(Fig. 12-5(a));

②By bearing on the member or bolt(Fig. 12-5(b));

③By tension in the member(Fig. 12-5(d));

④By shear at the end of the member(Fig. 12-5(c)).

The failure modes are shown in Fig. 12-5.

The failures noted above are prevented by taking the following measures.

①For modes (a) and (b), provide sufficient bolts of suitable diameter.

②For mode (c), design tension members for the effective area.

③Provide sufficient end distance for mode (d).

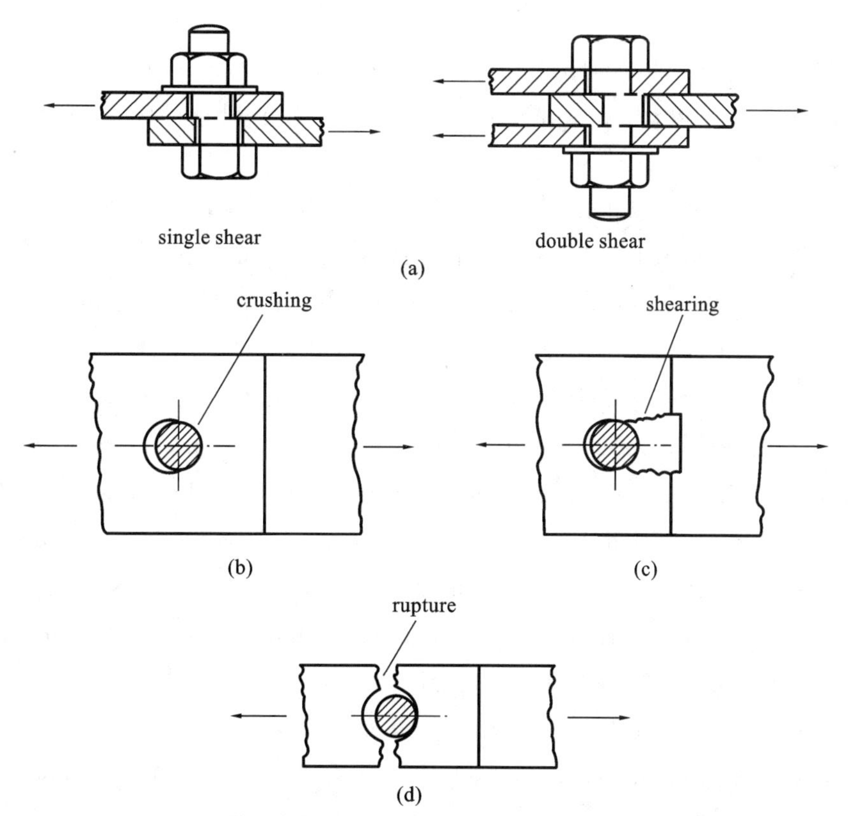

Fig. 12-5 Failure modes of a bolted joint

(a)shearing of bolt shank; (b)bearing of plate and bolt; (c)end shear failure; (d)tension failure of plate

Welded connections

(1)Welding.

Welding is the process of joining metal parts by fusing them and filling in with molten metal from the electrode. The method is used extensively to join parts and members, attach cleats, stiffeners, end plates, etc. and to fabricate complete elements such as plate girders. Welding produces neat, strong and more efficient joints than are possible with bolting. However, it should be carried out under close supervision, and this is possible in the fabrication shop. Site joints are usually bolted. Though site welding can be done it is costly, and *defects*(缺陷) are more likely to occur.

Electric arc welding(电弧焊) is the main system used, and the two main processes in structural steel welding as below.

①Manual arc welding, using a hand-held *electrode*(电焊条), coated with a flux which melts and protects the molten metal. The weld quality depends largely on the skill of the welder.

②Automatic arc welding. A continuous wire electrode is fed to the weld pool. The wire may be coated with flux or the flux can be supplied from a hopper. In another process an *inert gas*(惰性气体) is blown over the weld to give protection.

(2)Types of welds, defects and testing.

The two main types of welds, *butt*(对接焊缝) and *fillet*(角焊缝), are shown in Fig. 12-6(a) and (b). Butt welds are named after the edge preparation used. Single and double U and V welds are shown in Fig. 12-6(c). The double U weld requires less weld metal than the V type. A 90-degree fillet weld is shown but other angles are used. The weld size is specified by the leg length. Some other types of welds the partial butt, partial butt and fillet weld and deep penetration fillet weld—are shown in Fig. 12-6(d). In the *deep penetration fillet weld*(深熔透角焊缝) a higher current is used to fuse the plates beyond the limit of the weld metal.

Cracks can occur in welds and adjacent parts of the members being joined. Contraction on cooling causes cracking in the weld. Hydrogen absorption is the main cause of cracking in the heat-affected zone while *lamellar tearing*(层间撕裂) along a *slag inclusion*(夹渣)is the main problem in plates.

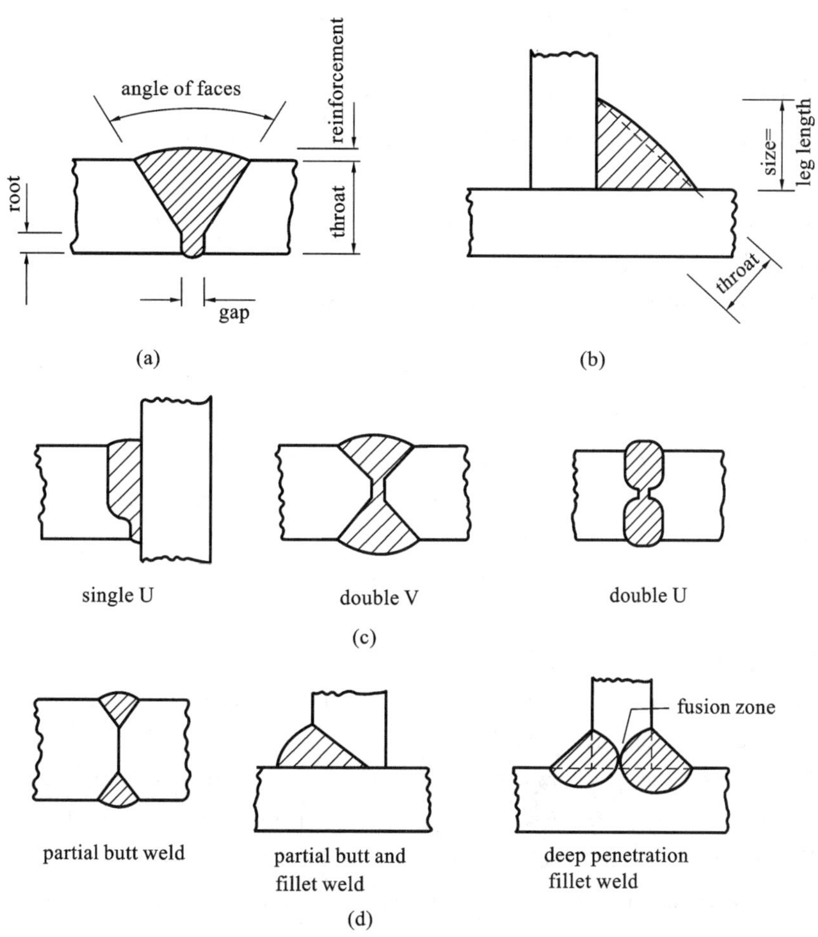

Fig. 12-6 Weld types: fillet and butt welds

(a) single V buff weld; (b) fillet weld; (c) types of buff welds; (d) other types of welds

土木工程专业英语论文写作技巧(3)——论文中的"比较"

在英语论文写作中,使用"比较"句可以揭示哪个更有用或更有价值。学会正确表达比较是很重要的。英语中最基本的用于表达"比较"的句型为比较级的用法。除此之外,还有其他常用于表达比较的句式。

(1)用于表达差异"differences"常用的句子。

- X is *different from* Y in a number of respects.
- There are a number of important *differences between* X and Y.
- X *differs from* Y in a number of important ways.
- Smith(2003) found distinct *differences between* X and Y.
- Women and men *differ not only in* physical attributes, *but also in* the way in which they...

• The *difference* in the deformation model assumed for an unstiffened connection and a stiffened connection is…

(2)用于表达相似性(similarities)常用的句子。

• The mode of processing used by the right brain *is similar to that* used by the left brain.

• The mode of processing used by the right brain *is comparable in* complexity to that used by the left brain.

• The development trend of the cracks was *similar to* the trend of the unstrengthened specimen.

• The effects of nitrous dioxide on human health *are similar to those of* ground level ozone.

• There are *a number of similarities between* X and Y.

• Numerous studies *have compared* the brain cells in humans and animals, showing that these cells *are essentially identical*.

• Therefore, a graph *similar to* Fig. 9 was obtained for the case of a long plate with…

(3)同一句中的比较(比较级)。

• Women are *faster/slower than* men at certain precision manual tasks, such as placing pegs in holes on a board.

• Women tend to perform *better/worse than* men on tests of perceptual speed.

• Adolescents are *less likely* to be put to sleep by alcohol *than* adults.

• Women tend to have *greater/less* verbal fluency *than* men.

• Men learned the route in *fewer* trials and made *fewer errors than* did women.

• It seems *more appropriate* to obtain the moment-rotation curves through the use of mathematical models.

除了比较级之外,其他用于表达比较"comparison"常用的句型。

• *In contrast to* oral communities, it is very difficult to get away from calendar time in literate societies.

• *Compared with* people in oral cultures, people in literate cultures organize their lives around clocks and calendars.

• Oral societies tend to be more concerned with the present, *whereas* literate societies have a very definite awareness of the past.

• Women's brains process language simultaneously on both sides of the brain, *while* men tend to process language on the left side only.

• This interpretation *contrasts with that* of Smith and Jones(2004), who argued that…

• *Compared to* closely spaced vertical columns in framed tube, diagrid structure consists of inclined columns on the exterior surface of building.

"compared to""compared with" 是可以代替 than 代表的对比句。但是它们二者又有区别。"compared to"用于表示两个不同事物的相似之处，而"compared with"用于表示两个相似事物的不同之处。例如在"*Compared with* externally bonded FRP materials, NSM systems are less prone to debonding from the concrete substrate and are less exposed to...(与外贴 FRP 材料相比，NSM 体系更不容易剥离也更不容易暴露于……)"中用于进行的对比双方 externally bonded FRP materials 与 NSM systems 是两个相似的事物。显然在英语论文写作中，需要使用"compared with"的情况更多些。

又例如下面两句：

The resulting orthogonality of the higher modes to the seismic input has the effect of greatly reducing the response of secondary equipment *as compared with* conventional design. 与常规设计相比，高阶模态与地震输入的正交性大大降低了次要设备的响应。

The newly introduced design elastic strain limit states are conservative and accurate *compared with* the current ultimate strain limit state.

(4)Comparison across two sentences(通过两句比较)。

• Load distribution in diagrid system is also studied for 36 storey building. *Similarly*, analysis and design of 50, 60, 70 and 80 storey diagrid structures is carried out.

上句中 similarly，可以根据不同句子的意思改为 by contrast，conversely 等。

(5)使用"比较句"时常见的错误——不完整的比较(incomplete comparisons)。

使用比较句时，要清楚地告诉读者而不是让读者去猜测比较对象。英语句子中要注意的是比较双方的"对等"，即动词跟动词、名词跟名词对比，永远不要把动词跟名词对比。一个不完整的比较句就会产生不对等的对比。例如在下句中：The seismic design specifications for the Wellington Police Station are considerably more severe than for the Union House in Auckland. Than 后面的介词短语"for the Union House in Auckland"是不可缺少的。

错误之一：中国学生最容易犯的错误为当对比的双方是动作的主语时，than 后面的主语往往缺少助动词。例如：The device formulated in this experiment exhibits higher luminance than conventional models *do*. 一句中的助动词"do"是最容易被忽略的。

一个不完整的比较句是会带来歧义的，例如在"Women prefer friendly doctors more than men"中，若读者根据字面意思，理解成如下两种意思都有可能："女性比男性更喜欢态度好的医生"或者"女性更喜欢态度好的医生，而不喜欢男医生"。因为该句从语法结构上没有指明 men 是跟 woman 对比还是跟 doctor 进行对比。但是若在完整的对比句"Women prefer friendly doctors more than men *do*"中，我们知道 men 是跟 woman 在对比。

错误之二:除了“do”经常被忽略外,“that”和“those”也经常被忽略。例如,在“Above the balls and attached to the building are cast-iron plates slightly concave similar to *those* below”一句中,those是经常被忽略的。再如,在下列完整的对比句“The flange strengthening scheme's reliability index was comparatively less than *that of* flange-web strengthening and closed-wrap strengthening”中,将几种strengthening scheme的reliability index进行了对比,而若将句子中的“that of”丢失,则从语法上表达的是将reliability index与后面的两种strengthening scheme进行对比。

错误之三:有些句子当中的介词“in”不能缺少。例如:Moreover, the floor slab may not be as continuous as in an in-situ build.

Lesson 13

STRUCTURAL SYSTEM OF CONCRETE BUILDING

Load transfer schemes

Reinforced concrete buildings are three-dimensional structures; however, they are generally constructed as an assemblage of more or less two-dimensional or planar subsystems lying primarily in the horizontal and vertical planes (e. g. floors, roofs, walls, plane frames, etc.). As an integrated system, the structure must carry and transfer to the foundation and the ground below all gravity loads on it as well as all horizontal loads and associated overturning moments resulting from wind or earthquake or asymmetry①. The structural system can, for the sake of convenience, be separated into gravity load resisting schemes and lateral load resisting schemes, although these two are mutually interacting and complementary.

Gravity load resisting schemes in building structures consist of a horizontal subsystem (floors and roofs) and a vertical subsystem (columns, walls, shafts, transfer girders, hangers, etc.).

Floor systems

The floor systems (horizontal subsystems) in building pick up the gravity loads acting on them and transfer them to the vertical elements (columns, walls, etc.). In this process, the floor system is primarily subjected to flexure and transverse shear and the vertical elements are subjected to axial force (usually compression), bending and shear. The floor also functions as a horizontal diaphragm connecting together and stiffening all the vertical elements. When these connections are rigid, "frame action" is also developed which greatly improves the efficiency of the system. In addition, the floor system, acting as diaphragms, picks up and distributes the horizontal loads generated in its tributary area (from wind or earthquake loadings) to the vertical elements of the main lateral load-resisting system. The diaphragms further help to maintain the overall cross-sectional geometry of the structure.

In cast-in-place reinforced concrete construction, the floor system usually consists of one of the following: ①slabs supported on bearing walls, ②flat plates, ③flat slabs, ④joist floors, ⑤beam and slab systems.

Wall supported slabs

A bearing wall and slab system is used mainly in low-rise residential type buildings. Slabs are generally 100 mm to 200 mm thick with spans from 3 m to 7.5

m. When the slab is supported only on opposite sides, the slab bends in one direction only and the slab is called a *one-way slab*. The entire load is transmitted one way to the two supports. A very similar situation exists in a long rectangular slab (length about twice greater than the width) supported on all four sides. The bending of the slab is predominantly along the short span and most of the load is transmitted across the short span onto the two longer supports. ② When the slab dimensions approach a square, the bending action along both spans becomes relatively significant. In this case, load is transmitted in both directions and the slab is termed a two-way slab.

Flat plates

A flat plate floor is a reinforced concrete slab of uniform thickness supported directly on columns (see Fig. 13-1). Edge beams are commonly added to stiffen exterior side frames which have heavy cladding loads. Flat plate floors are from 125 mm to 250 mm thick and are used with spans of 4.5 m to 6 m. Flat plate construction is suitable for such buildings as apartments and hotels where floor loads are small, long spans are not required, and where plate soffits can be used as ceilings. As the spans and/or loads increase, the shear and flexural stresses in the slab at the columns become very high. For moderate spans and loads, extra reinforcement may be used in this region, in the form of embedded steel beams or closely spaced bars. ③ However, for longer spans and heavier loads, the effective depth of the slab at the support regions has to be increased as is done in flat slab construction.

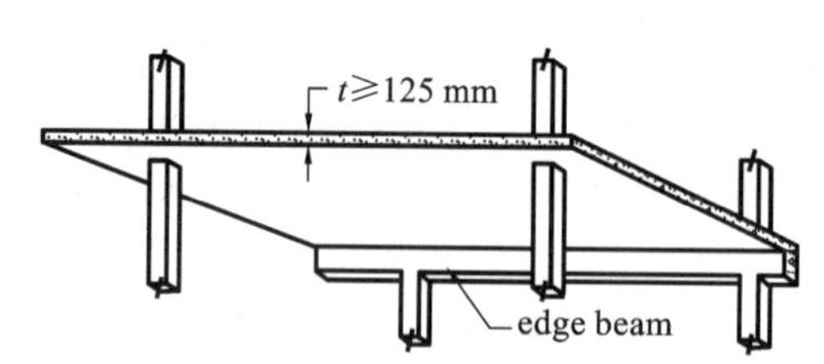

Fig. 13-1 Flat plate

Flat slabs

These are plates which are stiffened by drop panels and/or column capitals (see Fig. 13-2). Longer spans and heavier floor loads are possible with flat slab systems because of increased shear and negative moment capacity. Flat slabs are suitable for office construction when drop ceilings are provided and are used for spans up to about 10 m.

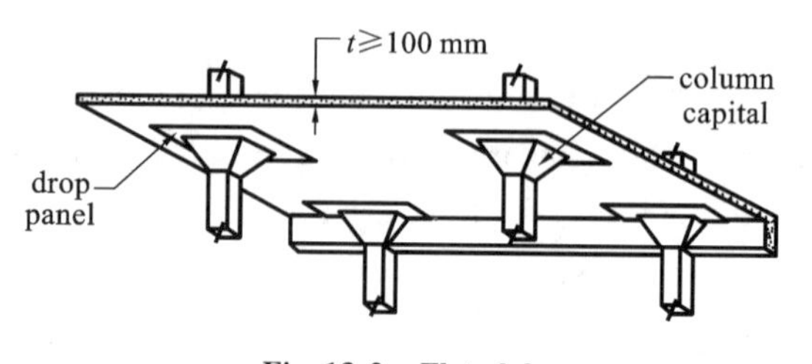

Fig. 13-2 Flat slab

Concrete joist floors

A concrete joist floor also called a ribbed slab, consist of a thin slab (usually 50 mm to 100 mm thick) cast integrally with relatively narrow, closely spaced stems (ribs) arranged in an one or two-way pattern. Ribs will have a thickness of 100 mm or more and a depth usually less than 3.5 times the minimum width. Two-way patterns are referred to as waffle slabs. Waffle slabs are built in 0.9 m to 1.5 m modules, span from 9.0 m to 12 m, and are about 300 mm to 600 mm deep overall. One-way joists use pans from 0.50 m to 0.80 m wide to form the ribs, and have span-depth ratios similar to those of waffle slabs.

Beam and slab

This type of construction consists of a slab built *monolithically* (整体的) with beams (girders) forming a grid pattern where the slab can be one-or two-way in nature depending on the dimensions of the slab panel (see Fig. 13-3). Slabs are commonly 100 mm to 150 mm thick and span 3 m to 7.5 m. Beam and slab construction is a light-weight system which adapts easily to floor openings and can be used for a wide range of column spacing.

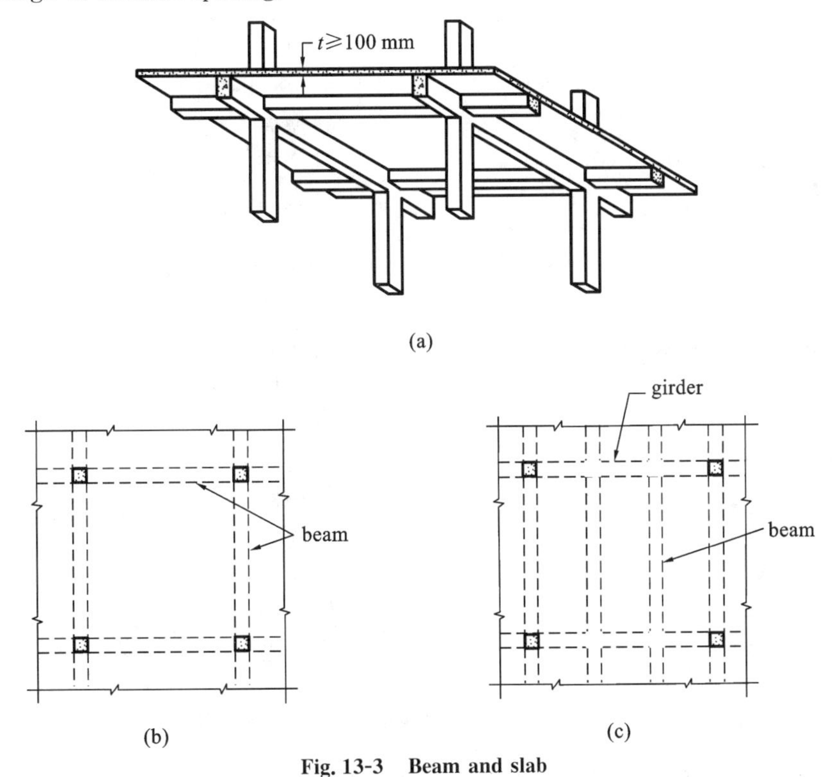

Fig. 13-3 Beam and slab

(a) two-way beam and slab; (b) two-way beam and slab; (c) one-way beam and slab

Methods have been developed to ensure that beams will have a proper safety margin against failure in flexure or shear, or due to inadequate bond and anchorage of the reinforcement. The member has been assumed to be at a hypothetical overload state for this purpose.

Words and Expressions

planar	[ˈpleinə]	*adj*. 平面的,平面内的
asymmetry	[əˈsimətri]	*n*. 不对称
shaft	[ʃɑːft]	*n*. 轴,支柱
girder	[ˈgəːdə]	*n*. 大梁
hanger	[ˈhæŋə]	*n*. 支架,托架
diaphragm	[ˈdaiəfræm]	*n*. 横隔板
tributary	[ˈtribjutəri]	*adj*. 附庸的,从属的,辅助的
cladding	[ˈklædiŋ]	*n*. 覆层
soffit	[ˈsɔfit]	*n*. 拱腹,下端背面
embed	[imˈbed]	*v*. 埋置,埋入,嵌进
rib	[rib]	*n*. 肋
module	[ˈmɔdjuːl]	*n*. 模数
pan	[pæn]	*n*. 底座
opening	[ˈəupəniŋ]	*n*. 开口,开洞
transfer girder		转换梁
cast-in-place reinforced concrete		现浇钢筋混凝土
bearing wall		承重墙
flat plate		平板
flat slab		(有柱帽的)无梁板
joist floor		密肋楼板
residential type building		居住型建筑
one-way slab		单向板
edge beam		边梁
flexural stress		弯曲应力
drop panel		平托板
column capital		柱帽
drop ceiling		吊顶
waffle slab		华夫板
span-depth ratio		跨高比

Notes

① As an integrated system, the structure must carry and transfer to the foundation and the ground below all gravity loads on it as well as all horizontal loads and associated overturning moments resulting from wind or earthquake or asymmetry.

该句主语为 the structure,谓语为 carry and transfer to,as well as 引导并列宾语 all gravity loads 和 all horizontal loads and associated overturning moments, resulting from wind or earthquake or asymmetry 为后置定语修饰 all horizontal loads and associated overturning moments。全句意思是：作为一个完整的体系，结构必须能承担并传递重力荷载、由风荷载或地震荷载或(结构)不对称引起的水平荷载和相关倾覆力矩到基础及其下的土体。

② The bending of the slab is predominantly along the short span and most of the load is transmitted across the short span onto the two longer supports.
该句的意思是：板的弯矩主要沿着短跨，荷载通过短跨传递到两个较长的支座上。

③ For moderate spans and loads, extra reinforcement may be used in this region, in the form of embedded steel beams or closely spaced bars.
句中 in the form of embedded steel beams or closely spaced bars 是介词短语作状语。全句的意思是：对中等跨度与中等荷载，埋置钢梁或密布钢筋形式的额外加固措施可能被用于这个区域。

Exercises

Ⅰ. Translate the following words into Chinese.

1. cast-in-place reinforced concrete ____________
2. bearing wall ____________
3. two-way slab ____________
4. flexure ____________
5. girder ____________

Ⅱ. Translate the following words into English.

1. 平板____________
2. 边梁____________
3. 吊顶____________
4. 柱帽____________
5. 墙支承板____________

Ⅲ. Translate the following sentences into Chinese.

1. In cast-in-place reinforced concrete construction, the floor system usually consists of one of the following: ①slabs supported on bearing walls, ②flat plates, ③flat slabs, ④joist floors, ⑤beam and slab systems.
2. Edge beams are commonly added to stiffen exterior side frames which have heavy cladding loads.
3. A concrete joist floor, also called a ribbed slab, consists of a thin slab (usually 50 mm to 100 mm thick) cast integrally with relatively narrow, closely spaced stems (ribs) arranged in an one or two-way pattern.
4. As the spans and/or loads increase, the shear and flexural stresses in the slab at the columns become very high.

5. Flat plate construction is suitable for such buildings as apartments and hotels where floor loads are small, long spans are not required, and where plate soffits can be used as ceilings.

Ⅳ. Translate the following sentences into English.

1. 当这些连接是刚性时,"框架作用"也被发挥,这能大大提高体系的效能。
2. 梁板型楼板是轻质体系,它易于适应楼板开洞并能在大范围的柱距中使用。
3. 为了方便起见,结构体系被分成重力荷载抵抗系统和水平荷载抵抗系统,且这两个系统互相作用,互为补充。
4. 建筑的楼板体系(水平子系统)承担作用其上的重力荷载,并传递给竖向构件(如柱、墙等)。
5. 如果板的跨度更大或板上荷载更大,支座处板的有效高度要像无梁楼板那样增大。

Ⅴ. Answer the following questions briefly.

1. What is the main function of the reinforced concrete structure of buildings?
2. Please tell the differences between the one-way slab and the two-way slab.
3. Why the beam and slab floor systems are widely used in building floor construction?

Reading Materials

Tall Building Structures and Design Process

Ideally, in the early stages of planning a building, the entire design team, including the architect, structural engineer, should collaborate to agree on a form of structure to satisfy their respective requirements of function, safety and serviceability, and servicing. A compromise between conflicting demands will be almost inevitable. In all but the very tallest structures, however, the structural arrangement will be *subservient*(辅助性的) to the architectural requirements of space arrangement and *aesthetics*(美学). Often, this will lead to a less-than-ideal structural solution that will *tax*(使负重担) the *ingenuity*(独创性), and probably the patience, of the structural engineer.

The two primary types of vertical load resisting elements of tall buildings are columns and walls. The latter acts either independently as shear walls or in assemblies as *shear wall cores*(芯筒剪力墙). The building function will lead naturally to the provision of walls to divide and *enclose*(包围,封隔)space, and of cores to contain and convey services such as elevators. Columns will be provided, in otherwise unsupported regions, to transmit gravity loads, and in some types of structure, horizontal loads also. Columns may also serve architecturally as, for example, facade *mullions*(竖框,直棂).

The inevitable primary function of the structural elements is to resist the gravity loading from the weight of the building and its contents. Since the loading on different floors tends to be similar, the weight of the floor system per unit floor area is approximately constant, regardless of the building height.

The highly probable second function of the vertical structural elements is also to resist the *parasitic*(附加的) load caused by wind and possibly earthquakes, whose magnitudes will be obtained from national building codes or wind tunnel studies. The bending moments on the building caused by these lateral forces increase with at least the square of the height, and their effects will become progressively more important as the building height increase. On the basis of the factors above, the relative quantities of material required in the floors, columns, and wind bracing of a traditional steel frame and the penalty on these due to increasing height are approximately.

Because the worst possible effects of lateral forces occur rarely, if ever, in the life of the building, it is *imperative*(必不可少的) to minimize the penalty for height to achieve an *optimum*(最优的) design. The constant search for more efficient solutions led to the innovative designs and new structural forms of recent years. In developing a suitable system for resisting lateral forces, the engineer seeks to devise stiff horizontal interconnections between the various vertical components to form composite assemblies such as coupled walls and rigid frames, which creates a total structural assembly having a lateral stiffness many times greater than the sum of the lateral stiffness of the individual vertical components.

The design process

Once the functional layout of the structure is decided, the design process generally follows a well-defined iterative procedure. Preliminary calculations for member sizes are usually based on gravity loading *augmented*(增大,扩大) by an *arbitrary*(任意的) *increment*(增加,增量) to account for wind forces. The cross sectional areas of the vertical member will be based on the accumulated loadings from their associated *tributary areas*(附属面积), with reductions to account for the probability that not all floors will be subjected simultaneously to their maximum live loading. The initial sizes of beams and slabs are normally based on moments and shears obtained from some simple methods of gravity load analysis, such as two-cycle *moment distribution*(弯矩分配), or from *codified*(编纂,整理) "mid-" and "-end" span values.

A check is then made on the maximum horizontal deflection, and the forces in the major structural members, using some rapid approximate analysis technique. If the deflection is *excessive*(过度的,过量的), or some of the members are inadequate, adjustments are made to the member sizes or the structural arrangement. If certain members attract excessive loads, the engineer may reduce their stiffness to redistribute the load to less heavily stressed components. The procedure of preliminary analysis, checking, and adjustment is repeated until a satisfactory solution is obtained.

Invariably, *alterations*(更替,变换) to the initial *layout*(布置) of the building will be required as the client's and architect's ideas of the building evolve. This will call for structural modifications, or perhaps a *radical*(根本的,主要的) rearrangement, which necessitates a complete review of the structural design. The various preliminary stages may therefore have to be repeated a number of times before a final solution is reached.

A *rigorous*(严格的,精确的) final analysis, using a more refined analytical model, will then be made to provide a final check on deflections and member strength. This will usually include the second-order effects of gravity loads on the lateral deflections and member forces, also called *P*-Delta effects(*P*-Δ 效应). A dynamic analysis may also be required if, as a result of wind loading, there is any likelihood of excessive deflections due to *oscillations*(振动,振荡) or of comfort criteria being exceeded, or if earthquake loading has to be considered. At some stage in the procedure the *deleterious*(有害的) effects of differential movements due to creep, shrinkage, or temperature *differentials*(差别) will also be checked.

土木工程专业英语论文写作技巧(4)——使用"强"动词

技术型作家喜欢强动词而非弱动词。强动词是一个表示动作的单词,如 compare、consider 和 indicate 等。弱动词或助动词需要一个以上的词来表示某动作,例如,make a comparison、give consideration to 和 give indication。作为学生,我们要使用强动词而不是弱动词。例如:

Smith *asks the questions as to* whether conventional method offer the best environment for execution.

Smith *questions* whether conventional methods offer the best environment for execution.

The column flange can *be idealized into a model* as a thin plate with width equal to Ho-2to.

The column flange can be *modelled* as a thin plate with width equal to Ho-2to.

在写作中常用的动词(词组)及其替代动词如下。

in some cases→occasionally

in this case→here

a number of→several/numerous

make clear→elucidate/clarify

make sure→endure/assure
meet→satisfy/fufill/adhere to
much/strong→markedly/considerably/substantially
realize→comprehend/perceive/understand
solve→alleviate/modify/resolve/eliminate/eradicate
suitable→appropriate/adequate
try→attempt/aim/aspire
usually→normally/typically/generally
very→highly/rather/quite/extremely
way→method/means/approach/strategy
whole→complete/entire/comprehensive
figure out→distinguish/differentiate/discriminate/identify
find→determine/derive/attain/locate/identify
help→assist/facilitate/guide/direct
important→critical/crucial/essential/pertinent/relevant/significant/vital
improve→enhance/upgrade/elevate
be made of→consist of/ comprises/be composed of
affect→influence/shape
carry out→implement/execute/promulgate/conduce
consider→evaluate/assess
check→verify/conform
different→distinct/diverse/various/varied
little/few→seldom/slightly
problem→limitation/restriction/obstacle/hindrance
need→require/stipulate/necessitate

Lesson 14

STIFFNESS AND DRIFT LIMITATIONS OF TALL BUILDING

The provision of adequate stiffness, particularly lateral stiffness, is a major consideration in the design of a tall building for several important reasons. As far as the ultimate limit state is concerned, lateral deflections must be limited to prevent second order *P*-Delta effects due to gravity loading being of such a magnitude as to participate collapse. ① In terms of the serviceability limit state, first, deflections must be maintained at a sufficiently low level to allow the proper functioning of nonstructural components such as elevators and doors; second, to avoid distress in the structure, to prevent excessive cracking and consequent loss of stiffness, and to avoid any redistribution of load to non-load-bearing partitions, infills, claddings, or glazing; and third, the structure must be sufficiently stiff to prevent dynamic motions becoming large enough to cause discomfort to occupants, prevent delicate work being undertaken, or affect sensitive equipment. In fact, it is in the particular need for concern for the provision of lateral stiffness that the design of a high-rise building largely departs from that of a low-rise building.

One simple parameter that affords an estimate of the lateral stiffness of a building is the drift index, defined as the ratio of the maximum deflection at the top of the building to the total height. In addition, the corresponding value for a single story height, the interstory drift index, gives a measure of possible localized excessive deformation. The control of lateral deflection of particular importance for modern buildings in which the traditional reserves of stiffness due to heavy internal partitions and outer cladding has largely disappeared. It must be stressed, however, that even if the drift index is kept within the traditionally accepted limits, such as 1/500, it does not necessarily follow that the dynamic comfort criteria will also be satisfactory. Problems may arise, for example, if there is coupling between bending and torsional oscillations that lead to unacceptable complex motions or accelerations. ② In addition to static deflection calculations, the question of the dynamic response, involving the lateral acceleration, amplitude, and period of oscillation, may also have to be considered.

The establishment of a drift index limit is a major design decision, but, unfortunately, there are no unambiguous or widely accepted values, or even, in some of the national codes concerned, any firm guidance. The designer is then faced with having to decide on an appropriate value. The figure adopted will reflect the building usage, the type of design criterion employed (for example, working or ultimate load conditions), the form of construction, the materials employed, including any substantial infills or claddings the wind loads considered, and, in particular, past experience of similar buildings that have performed satisfactorily.

Design drift index limits that have been used in different countries range from 0.001 to 0.005. To put this in perspective, a maximum horizontal top deflection of between 0.1 and 0.5 m (6 to 20 in.) would be allowed in a 33 story, 100 m (330 ft.) high building, or, alternatively, a relative deflection of 3 to 15 mm (0.12 to 0.6 in.) over a story height of 3 m (10 ft.). Generally, lower values should be used for hotels or apartment buildings, since noise and movement tend to be more disturbing in the former. Consideration may be given to whether the stiffening effects of any internal partitions, infills, or claddings are included in the deflection calculations.

The consideration of this limit state requires an accurate estimate of the lateral deflections that occur, and involves an assessment of the stiffness of cracked members, the effects of shrinkage and creep and any redistribution of forces that may result, and of any rotational foundation movement. In the design process, the stiffness of joints, particularly in precast or prefabricated structures, must be given special attention to develop adequate lateral stiffness of the structure and to prevent any possible progressive failure. The possibility of torsional deformations must not be overlooked.

In practice, non-load-bearing infills, partitions, external wall panels, and window glazing should be designed with sufficient clearance or with flexible supports to accommodate the calculated movements.

Sound engineering judgment is required when deciding on the drift index limit to be imposed. However, for conventional structures, the preferred acceptable range is 0.0015 to 0.003 (that is, approximately 1/650 to 1/350), and sufficient stiffness must be provided to ensure that the top deflection does not exceed this value under extreme load conditions. As the height of the building increases, drift index coefficients should be decreased to the lower end of the range to keep the top story deflection to a suitably low level. ③

The drift criteria apply essentially to quasi-static conditions. When extreme force conditions are possible, or where problems involving vortex shedding or other unusual phenomena may occur, a more sophisticated approach involving a dynamic analysis may be required.

If excessive, the drift of the structure can be reduced by changing the geometric configurations to alter the mode of lateral load resistance, increasing the bending stiffness of the horizontal members, adding additional stiffness by the inclusion of stiffer wall or core members, achieving stiffer connections, and even by slopping the exterior columns. In extreme circumstances, it may be necessary to add dampers, which may be of the passive or active type.

Words and Expressions

drift	[drift]	*n.* 层间侧移
		v. 侧移
partition	[pɑːˈtiʃən]	*n.* 隔墙
infill	[ˈinfil]	*n.* 填充物,填补空间
		v. 填充
cladding	[ˈklædiŋ]	*n.* 包层
glazing	[ˈgleiziŋ]	*n.* 窗玻璃
occupant	[ˈɔkjupənt]	*n.* 居住者,占有者
parameter	[pəˈræmitə]	*n.* 参数,系数
oscillation	[ɔsiˈleiʃən]	*n.* 摆动,振动
complex	[ˈkɔmpleks]	*adj.* 复合的,综合的
amplitude	[ˈæmplitjuːd]	*n.* 振幅,幅度
unambiguous	[ˈʌnæmˈbigjuəs]	*adj.* 明确的,不含糊的
prefabricate	[ˌpriːˈfæbrikeit]	*v.* 预制,装配
clearance	[ˈkliərəns]	*n.* 间隙,净空
sound	[saund]	*adj.* 健全的,可靠的,合理的
vortex	[ˈvɔːteks]	*n.* 涡流,旋涡
damper	[ˈdæmpə]	*n.* 阻尼器
lateral stiffness		侧向刚度,抗侧刚度
P-Delta effect		P-Δ 效应
interstory drift index		层间侧移指数(限值)
in perspective		恰当地,正确地
quasi-static		伪静力
vortex shedding		涡流脱落

Notes

① As far as the ultimate limit state is concerned, lateral deflections must be limited to prevent second order *P*-Delta effects due to gravity loading being of such a magnitude as to participate collapse.
就极限状态而言,必须限制侧向变形以防止重力荷载引起的二阶 P-Δ 效应,(当二阶效应)较大时,将参与结构倒塌。

② Problems may arise, for example, if there is coupling between bending and torsional oscillations that lead to unacceptable complex motions or accelerations.

例如当弯扭振动耦合引起不能接受的复合运动或加速度时就会出现问题。

③ As the height of the building increases, drift index coefficients should be decreased to the lower end of the range to keep the top story deflection to a suitably low level.

随着建筑物高度的增加,侧移指数应减小至可取范围内较小的值以使顶层变形保持在一个合适的较低水平。

Exercises

Ⅰ. Translate the following words into Chinese.

1. serviceability limit state ____________________
2. drift index ____________________
3. dynamic response ____________________
4. prefabricated structures ____________________
5. bending stiffness ____________________

Ⅱ. Translate the following words into English.

1. 层间侧移____________________
2. 低层建筑____________________
3. 极限状态____________________
4. 抗侧刚度____________________

Ⅲ. Translate the following sentences into Chinese.

1. One simple parameter that affords an estimate of the lateral stiffness of a building is the drift index, defined as the ratio of the maximum deflection at the top of the building to the total height.
2. The figure adopted will reflect the building usage, the type of design criterion employed (for example, working or ultimate load conditions), the form of construction, the materials employed, including any substantial infills or claddings the wind loads considered, and, in particular, past experience of similar buildings that have performed satisfactorily.
3. The consideration of this limit state requires an accurate estimate of the lateral deflections that occur, and involves an assessment of the stiffness of cracked members, the effects of shrinkage and creep and any redistribution of forces that may result, and of any rotational foundation movement.
4. If excessive, the drift of the structure can be reduced by changing the geometric configurations to alter the mode of lateral load resistance, increasing the bending stiffness of the horizontal members, adding additional stiffness by the inclusion of stiffer wall or core members, achieving stiffer connections,

and even by sloping the exterior columns.

Ⅳ. Translate the following sentences into English.

1. 高层建筑结构必须具有足够的侧向刚度以保证其使用安全。
2. 侧移过大会使高层建筑出现变形破坏,因而应对此加以限制。
3. 结构的侧移可通过改变其几何构成、增加构件刚度、利用加劲墙和核心构件来减小。
4. 在特定条件下,在结构中增加阻尼器也是减小侧移的有效方法。

Ⅴ. Answer the following questions briefly.

1. Why can the provision of adequate stiffness be a major consideration in the design of a tall building?
2. What is drift index?
3. For what reasons are the traditionally accepted values of drift index not satisfactory?
4. In practice, how can a designer decide an appropriate value of drift index?
5. By what means can drift index be reduced?

Reading Materials

Structural Forms of Tall Buildings

A major step forward in reinforced concrete high-rise structural form comes with the introduction of shear walls for resisting horizontal loading. This is the first in a series of significant developments in the structural forms of concrete high-rise buildings, freeing them from the previous 20 to 25 story height limitations of the rigid-frame and flat plate systems. The innovation and *refinement*(精致) of these new forms, together with the development of higher strength concretes, has allowed the height of concrete buildings to reach within striking distance of 100 stories.

Of the following structural forms, some are more appropriate to steel and others to reinforced concrete; many are suitable for either material, while a few allow or demand a combination of materials in the same structure. They are described in a roughly historical sequence.

***Braced-frame*(框架-支撑)structures**

In braced frames the lateral resistance of the structure is provided by *diagonal*(斜的,对角线的) members that, together with the girders, form the "web" of the vertical truss, with the columns acting as the "chords". Because the horizontal shear on the building is resisted by the horizontal components of the axial tensile or compressive actions in the web members, bracing systems are highly efficient in resisting lateral loads.

The traditional use of bracing has been in story-height, bay-width modules (see Fig. 14-1) that are fully concealed in the finished building. More recently, however, external larger scale bracing, extending over many stories and *bays*(开间,跨), has been used to produce not only highly efficient structures, but aesthetically attractive buildings (see Fig. 14-2).

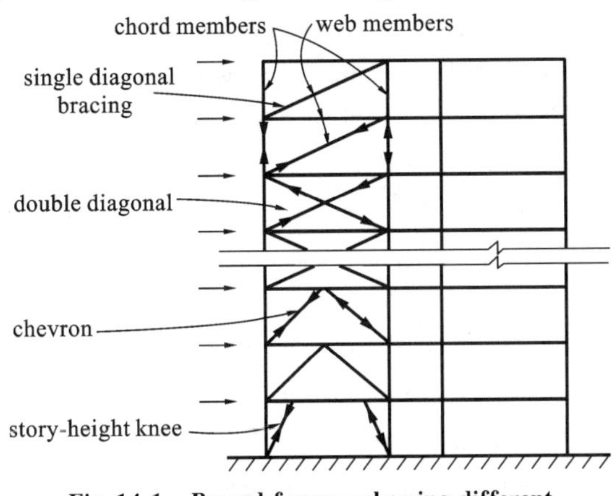

Fig. 14-1 Braced frame—showing different types of bracing

Fig. 14-2 Bank of China in Hong Kong

***Rigid-frame*(刚架)structures**

Rigid-frame structures consist of columns and girders joined by moment-resistant connections. The lateral stiffness of a rigid frame bent depends on the bending stiffness of the columns, girders, and connections in the plane of the bent. The rigid frame's principal advantage is its open *rectangular*(矩形的) arrangement, which allows freedom of planning and easy fitting of doors and windows. If used as the only source of lateral resistance in a building, in its typical 20 ft (6 m)-30 ft (9 m) bay size, rigid framing is economic only for buildings up to about 25 stories. Above 25 stories the relatively high lateral flexibility of the frame calls for uneconomically large members in order to control the drift.

Gravity loading also is resisted by the rigid-frame action. Negative moments are induced in the girders adjacent to the columns causing the mid-span positive moments to be significantly less than in a simply supported span. In structures that gravity loads dictate the design, economies in member sizes that arise from this effect tend to be *offset*(抵消) by the higher cost of the rigid joints.

***Flat-plate and flat-slab*(无梁楼板)structures**

The flat-plate structure is the simplest and most logical of all structural forms in that it consists of uniform slabs, of 5-8 in. (12-20 cm) thickness, connected rigidly to supporting columns. The system, which is essentially of reinforced concrete, is very economical in having a flat soffit requiring the most uncomplicated formwork and, because the soffit can be used as the ceiling, in creating a minimum possible *floor depth*(板厚).

***Shear wall*(剪力墙)structures**

Concrete or masonry continuous vertical walls may serve both architecturally as partitions and structurally to carry gravity and lateral loading. Their very high *in plane*(平面内) stiffness and strength make them ideally suited for bracing tall buildings. In a shear wall structure, such walls are entirely responsible for the lateral load resistance of the building. They act as vertical cantilevers in the form of separate planar walls, and as nonplanar assemblies of connected walls around elevator, stair, and service *shafts*(竖井,通风井). Because they are much stiffer horizontally than rigid frames, shear wall structures can be economical up to about 35 stories.

In contrast to rigid frames, the shear walls' solid form tends to restrict *planning*(平面布置) where open internal spaces are required. They are well suited, however, to hotels and residential buildings where the floor-by-floor repetitive planning allows the walls to be vertically continuous and where they serve simultaneously as excellent *acoustic*(声学的) and fire *insulators*(分隔墙) between rooms and apartments.

***Coupled wall*(联肢剪力墙)structures**

A coupled wall structure is a particular, but very common, form of shear wall structure with its own special problems of analysis and design. It consists of two or more shear walls in the same plane, or almost the same plane, connected at the floor levels by beams or stiff slabs. The effect of the shear-resistant connecting members is to cause the set of walls to behave in their plane partly as a composite cantilever, bending about the common *centroidal*(重心的) axis of the walls. This results in a horizontal stiffness very much greater than if the walls acted as a set of separate uncoupled cantilevers.

***Wall-frame*(框架-剪力墙) structures**

When shear walls are combined with rigid frames, the walls, which tend to deflect in a flexural configuration, and the frames, which tend to deflect in a shear mode, are constrained to adopt a common deflected shape by the horizontal rigidity of the girders and slabs. As a consequence, the walls and frames interact horizontally, especially at the top, to produce a stiffer and stronger structure. The interacting wall-frame combination is appropriate for buildings in the 40 to 60 story range, well beyond that of rigid frames or shear walls alone.

***Framed-tube*(框筒)structures**

The lateral resistance of framed-tube structures is provided by very stiff moment-resisting frames that form a "tube" around the perimeter of the building. The frames consist of closely spaced columns, 6-12 ft (2-4 m) between centers, joined by deep spandrel girder. Although the tube carries all the lateral loading, the gravity loading is shared between the tube and interior columns or walls. When lateral loading acts, the perimeter frames *aligned*(对准,对齐) in the direction of loading act as the "webs" of the massive tube cantilever, and those *normal to*(垂直于) the direction of the loading act as the "flanges".

The close spacing of the columns throughout the height of the structure is usually unacceptable at the entrance level. The columns are therefore merged, or terminated on a *transfer beam*(转换梁), a few stories above the base so that only a few, larger, more widely spaced columns continue to the base. The tube form is developed originally for buildings of rectangular plan, and probably its most efficient use is in that shape. It is appropriate, however, for other plan shapes, and has occasionally been used in circular and triangular configurations.

The tube is suitable for both steel and reinforced concrete construction and used for buildings ranging from 40 to more than 100 stories.

The tube structure's structural efficiency, although high, still leaves scope for improvement because the "flange" frames tend to suffer from "*shear lag*"(剪力滞); this results in the mid-face "flange" columns being less stressed than the corner columns and, therefore, not contributing as fully as they can to the flange action.

***Tube-in-tube*(筒中筒)or *hull-core*(筒体)structures**

This variation of the framed tube consists of an outer framed tube, the "hull" together with an internal elevator and service core. The hull and core act jointly in resisting both gravity and lateral loading. In a steel structure the core may consist of braced frames, whereas in a concrete structure it would consist of an assembly of shear walls.

To some extent, the outer framed tube and the inner core interact horizontally as the shear and flexural components of a wall-frame structure, with the benefit of increased lateral stiffness. However, the structural tube usually adopts a highly dominant role because of its much greater structural depth.

***Bundled-tube*(成束筒)structures**

This structural form is notable in its having been used for the Sears Tower in Chicago (see Fig. 14-3). The tower consists of four parallel rigid steel frames in each orthogonal direction,

Fig. 14-3 Sears Tower in Chicago

interconnected to form nine "bundled" tubes. As in the single-tube structure, the frames in the direction of lateral loading serve as "webs" of the vertical cantilever, with the normal frames acting as "flanges".

土木工程专业英语论文写作技巧(5)——避免在句首使用 it

在口语和非正式写作中,可以使用代词"it"和 "there"作形式主语。但是在正式写作中,不应该经常使用这种形式,以免语句冗长。当然,读者可以在避免同一个句型使用过多而作出改变时使用。

本科和研究生写作水平的一个关键区别是学生在书面项目中使用的单词数量。本科生试图用尽可能多的单词,因为他们用于描述一个项目至少有 500 个单词。强调数量胜过质量不利于专业英语写作。技术型作家强调表达想法,而不是他们的单词数量,所以使用的句子尽可能清楚和简短。他们希望论文是由简单的句子结构组成,而不是由大量字数填满他们的论文,这样就更能突出研究结果。

There is a necessity for a semi-structured approach to be chosen. →A semi-structured approach *must be chosen*.

It is important to establish appropriate models to predict the deformation of the individual connection elements. →Appropriate models to predict the deformation of the individual connection elements *must be established*.

It is essential that the model be revised. →The model *must be revised*.

It is necessary to predict the connection response accurately. →An accurate prediction of the connection response *is necessary*. /The connection response *must be predicted* accurately.

下列句子允许真正的主体被放置在一个清晰的位置,同时创造更直接的谓语。为避免句子过于冗长,而是用短语以"it"开头。

It is known that…

It is considered that…

It demonstrates that…

It could be said that…

It follows that…

Lesson 15

SOURCES OF PRESTRESS FORCE

Prestress can be applied to a concrete member in many ways. Perhaps the most obvious method of pre-compressing is to use jacks reacting against abutments, as shown in Fig. 15-1(a). Such a scheme has been employed for large projects. Many variations are possible, including replacing the jacks with compression struts after the desired stress in the concrete is obtained or using inexpensive jacks that remain in place in the structure, in some cases with a cement grout used as the hydraulic fluid. The principal difficulty associated with such a system is that even a slight movement of the abutments will drastically reduce the prestress force.

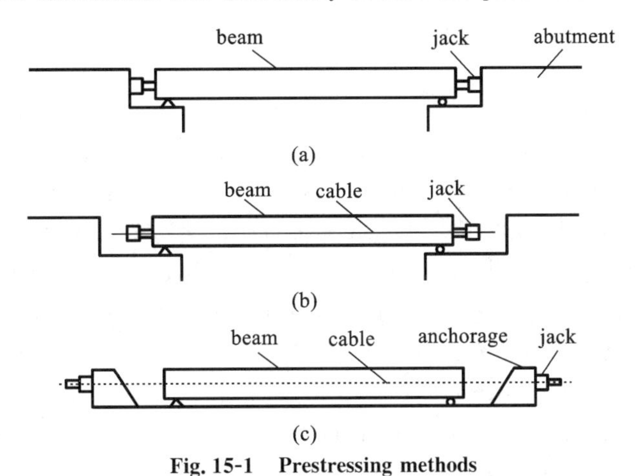

Fig. 15-1 Prestressing methods

(a)post-tensioning by jacking against abutments; (b)post-tensioning with jacks reacting against beam; (c)pretensioning with tendon stressed between fixed external anchorages

In most cases, the same result is more conveniently obtained by tying the jack bases together with wires or cables, as shown in Fig. 15-1(b). These wires or cables may be external, located on each side of the beam; more usually they are passed through a hollow conduit embedded in the concrete beam. Usually, one end of the prestressing tendon is anchored, and all of the force is applied at the other end. After reaching the desired prestress force, the tendon is wedged against the concrete and the jacking equipment is removed for reuse. ①In this type of prestressing, the entire system is self-contained and is independent of relative displacement of the supports.

Another method of prestressing that is widely used is illustrated in Fig. 15-1 (c). The prestressing strands are tensioned between massive abutments in a casting yard prior to placing the concrete in the beam forms. The concrete is placed around the tensioned strands, and after the concrete has attained sufficient strength, the jacking pressure is released. This transfers the prestressing force to the concrete by bond and friction along the strands, chiefly at the outer ends.

It is essential, in all three cases shown in Fig. 15-1, that the beam be supported in such a way as to permit the member to shorten axially without restraint so that the prestressing force can be transferred to the concrete. ②

Other means for introducing the desired prestressing force have been attempted on an experimental basis. Thermal prestressing can be achieved by preheating the steel by electrical or other means. Anchored against the ends of the concrete beam while in the extended state, the steel cools and tends to contract. The prestress force is developed through the restrained contraction. The use of expanding cement in concrete members has been tried with varying success. The volumetric expansion, restrained by steel strands or by fixed abutments, produces the prestress force.

Most of the patented systems for applying prestress in current use are variations of those shown in Fig. 15-1(b) and (c). Such systems can generally be classified as pretensioning or post-tensioning systems. In the case of pretensioning, the tendons are stressed before the concrete is placed, as in Fig. 15-1(c). This system is well-suited for mass production, since casting beds can be made several hundred feet long, the entire length cast at once, and individual beams can be fabricated to the desired length in a single casting. ③

In post-tensioned construction, shown in Fig. 15-1(b), the tendons are tensioned after the concrete is placed and has gained its strength. Usually, a hollow conduit or sleeve is provided in the beam, through which the tendon is passed. ④ In some cases, tendons are placed in the interior of hollow box-section beams. The jacking force is usually applied against the ends of the hardened concrete, eliminating the need for massive abutments. In Fig. 15-2, six tendons, each consisting of many individual strands, are being post-tensioned sequentially using a portable hydraulic jack.

A large number of particular systems, steel elements, jacks, and anchorage fittings have been developed in this country and abroad, many of which differ from each other only in minor details. As far as the designer of prestressed concrete structures is concerned, it is unnecessary and perhaps even undesirable to specify in detail the technique that is to be followed and the equipment to be used. It is frequently best to specify only the magnitude and line of action of the prestress force. The contractor is then free, in bidding the work, to receive quotations from several

Fig. 15-2 Using a portable jack to stress multistrand tendons

different prestressing subcontractors, with resultant cost savings. It is evident, however, that the designer must have some knowledge of the details of the various systems contemplated for use, so that in selecting cross-sectional dimensions, any one of several systems can be accommodated.

Words and Expressions

prestress	[ˈpriːˈstres]	*v.* 对……预加应力 *n.* 预应力
jack	[dʒæk]	*n.* 千斤顶 *v.* 用千斤顶托起
abutment	[əˈbʌtmənt]	*n.* 墩台,桥台
scheme	[skiːm]	*n.* 方案,计划
tendon	[ˈtendən]	*n.* 预应力钢筋束(钢丝束)
wire	[ˈwaiə]	*n.* 钢丝,钢丝索
cable	[ˈkeibl]	*n.* 缆,索,钢丝绳,绳索
conduit	[ˈkɔndit]	*n.* 导管,管道
anchor	[ˈæŋkə]	*v.* 锚固 *n.* 锚具
wedge	[wedʒ]	*n.* 楔 *v.* 楔入,楔进
strand	[strænd]	*n.* 股,绞
patent	[ˈpeitənt, ˈpætənt]	*v.* 获得……的专利权 *n.* 专利
sleeve	[sliːv]	*n.* 套筒
fitting	[ˈfitiŋ]	*n.* 配件,装置
quotation	[kwəuˈteiʃən]	*n.* 报价

subcontractor [ˌsʌbkənˈtræktə(r)] n. 分包商
cement grout 水泥浆
hydraulic fluid 液压油
post-tensioning 后张法
pretensioning 先张法
mass production 批量生产

Notes

① In most cases, the same result is more conveniently obtained by tying the jack bases together with wires or cables, as shown in Fig. 15-1(b). These wires or cables may be external, located on each side of the beam; more usually they are passed through a hollow conduit embedded in the concrete beam. Usually, one end of the prestressing tendon is anchored, and all of the force is applied at the other end. After reaching the desired prestress force, the tendon is wedged against the concrete and the jacking equipment is removed for reuse.
本段简要介绍了常用的施加预应力的方式。本段意思为——大多数情况下,将千斤顶底座和钢丝或钢索系在一起,可以更方便地获得同样的结果,如图 15-1(b)所示。这些钢丝或钢索可以放在梁的外边两侧,更常见的是使它穿过埋设在混凝土梁内的空心导管。通常,将预应力钢束的一端锚固,在另一端施加全部力。当达到要求的预应力值后,将钢束楔紧在混凝土中,千斤顶设备可以移走以便再次使用。

② It is essential, in all three cases shown in Fig. 15-1, that the beam be supported in such a way as to permit the member to shorten axially without restraint so that the prestressing force can be transferred to the concrete.
本句中,第一个 that 引出一个表语从句。本句意思为——在如图 15-1 所示的所有三种情况下,梁的支承条件必须允许其毫无约束地轴向缩短,以便将预加力传递至混凝土。

③ In the case of pretensioning, the tendons are stressed before the concrete is placed, as in Fig. 15-1(c). This system is well-suited for mass production, since casting beds can be made several hundred feet long, the entire length cast at once, and individual beams can be fabricated to the desired length in a single casting.
这两句简要介绍了先张法的主要特点。这两句意思为——对于先张法,钢束在混凝土浇筑之前施加应力,如图 15-1(c)所示。这种体系非常适合大量生产,浇筑长度可长达数百英尺,沿全长一次浇筑混凝土,再把长的浇筑体分割成所需长度的单个构件。

④ In post-tensioned construction, shown in Fig. 15-1(b), the tendons are tensioned after the concrete is placed and has gained its strength. Usually, a hollow conduit or sleeve is provided in the beam, through which the tendon is passed.

这两句简要介绍了后张法的主要特点。这两句意思为——后张法施工在混凝土浇筑并达到一定强度后才张拉钢束，如图 15-1(b)所示。通常，梁内有空心导管或套筒，钢束可从其中穿过去。

Exercises

Ⅰ. Translate the following words into Chinese.

1. anchorage fittings ________________
2. the restrained contraction ________________
3. the magnitude and line of action of the prestress force ________________

Ⅱ. Translate the following words into English.

1. 空心导管________________
2. 箱形截面梁________________

Ⅲ. Translate the following sentences into Chinese.

1. Prestress can be applied to a concrete member in many ways. Perhaps the most obvious method of pre-compressing is to use jacks reacting against abutments.
2. Anchored against the ends of the concrete beam while in the extended state, the steel cools and tends to contract.
3. A large number of particular systems, steel elements, jacks, and anchorage fittings have been developed in this country and abroad, many of which differ from each other only in minor details.

Ⅳ. Translate the following sentences into English.

1. 先张法预应力构件中，将预张力传递给混凝土的方法是靠预应力筋，特别是两端部分的黏结作用和摩擦作用。
2. 后张法预应力施工时，千斤顶的力一般作用在已硬化混凝土构件的端部。
3. 设计者在设计时，必须对不同的预应力体系的细节有一定的了解。

Ⅴ. Answer the following questions briefly.

1. What is the major difference between pretensioning and post-tensioning?
2. What is the function of massive abutments in pretensioning construction?

Reading Materials

Introduction of Prestressed Concrete

Modern structural engineering tends to progress toward more economical structures through gradually improved methods of design and the use of higher-strength materials. This results in a reduction of cross-sectional dimensions and consequent weight savings. Such developments are particularly important in the field of reinforced concrete, where the dead load represents a substantial part of the total load. Also, in *multistory*(多层的) buildings, any saving in depth of members, multiplied by the number of stories, can represent a substantial saving in total height, load on foundations, length of heating and electrical ducts, plumbing *risers*(竖管), and wall and partition surfaces.

Significant savings can be achieved by using high-strength concrete and steel in conjunction with *present-day*(现代的) design methods, which permit an accurate *appraisal*(估计) of member strength. However, there are limitations to this development, mainly due to the interrelated problems of cracking and deflection at service loads. The efficient use of high-strength steel is limited by the fact that the amount of cracking (width and number of cracks) is proportional to the strain, and therefore the stress, in the steel. Although a moderate amount of cracking is normally not objectionable in structural concrete, excessive cracking is undesirable in that it exposes the reinforcement to corrosion: It may be visually offensive, and it may *trigger*(触发,导致) a premature failure by diagonal tension. The use of high strength materials is further limited by deflection considerations, particularly when refined analysis is used. The slender members that result may permit deflections that are functionally or visually unacceptable. This is further *aggravated*(恶化) by cracking, which reduces the flexural stiffness of members.

These limiting features of ordinary reinforced concrete have been largely overcome by the development of prestressed concrete. A prestressed concrete member can be defined as one in which there have been introduced internal stresses of such magnitude and distribution that the stresses resulting from the given external loading are counteracted to a desired degree. Concrete is basically a compressive material, with its strength in tension being relatively low. Prestressing applies a pre-compression to the member that reduces or eliminates undesirable tensile stresses that would otherwise be present. Cracking under service loads can be minimized or even avoided entirely. Deflections may be limited to an acceptable value; in fact, members can be designed to have zero deflection under the combined effects of service load and prestress force. Deflection and crack control, achieved through prestressing, permit the engineer to make use of efficient and economical high-strength steels in the form of strands, wires, or bars, in conjunction with concretes of much higher strength than normal. Thus, prestressing results in the overall improvement in performance of structural concrete used for ordinary loads and spans and extends the range of application far beyond the limits for ordinary reinforced concrete, leading not only to much longer spans than previously thought possible, but permitting innovative new structural forms to be employed.

土木工程专业英语论文写作技巧(6)——描述图形和表格的句型

1. { Table 5 / Fig. 5 } + { shows / compares / presents / provides / illustrates } + { the experimental data on x / the results obtained from a preliminary analysis of x / the intercorrelations among the nine measurements of x / some of the main characteristics of the… / the breakdown of… }

   - Table 5
   - Fig. 5

   \+
   - shows
   - compares
   - presents
   - provides
   - illustrates

   \+
   - the experimental data on x
   - the results obtained from a preliminary analysis of x
   - the intercorrelations among the nine measurements of x
   - some of the main characteristics of the…
   - the breakdown of…

2. { The results obtained from the can be seen preliminary analysis of x } + { are shown / can be seen / are compare / are presented } + { inTable 5 / in Fig. 5 }

   - The results obtained from the can be
     seen preliminary analysis of x

   \+
   - are shown
   - can be seen
   - are compare
   - are presented

   \+
   - inTable 5
   - in Fig. 5

3. { As shown in Fig. 1 / As can be seen from Table 1 / It can be seen from the data in Table 1 that / From Fig. 1, we can see that } + { the x group reported significantly more y than the other two groups }

   - As shown in Fig. 1
   - As can be seen from Table 1
   - It can be seen from the data
     in Table 1 that
   - From Fig. 1, we can see that

   \+
   - the x group reported significantly
     more y than
     the other two groups

Lesson 16

PRESTRESSING STEELS

Early attempts at prestressing concrete were unsuccessful because steel with ordinary structural strength was used. The low prestress obtainable in such rods was quickly lost due to shrinkage and creep in the concrete. ①

Such changes in length of concrete have much less effect on prestress force if that force is obtained using highly stressed steel wires or cables. In Fig. 16-1(a), a concrete member of length L is prestressed using steel bars with ordinary strength stressed to 24 000 psi. With $E_s = 2.9 \times 10^7$, the unit strain ε_s required to produce the desired stress in the steel of 24 000 psi is

$$\varepsilon_s = \frac{\Delta L}{L} = \frac{f_s}{E_s} = \frac{24\ 000}{2.9 \times 10^7} = 8.0 \times 10^{-4}\ \varepsilon \qquad 16\text{-}1$$

However, the long-term strain in the concrete due to shrinkage and creep alone, if the prestress force were maintained over a long period, would be on the order of 8.0×10^{-4} and would be sufficient to completely relieve the steel of all stress.

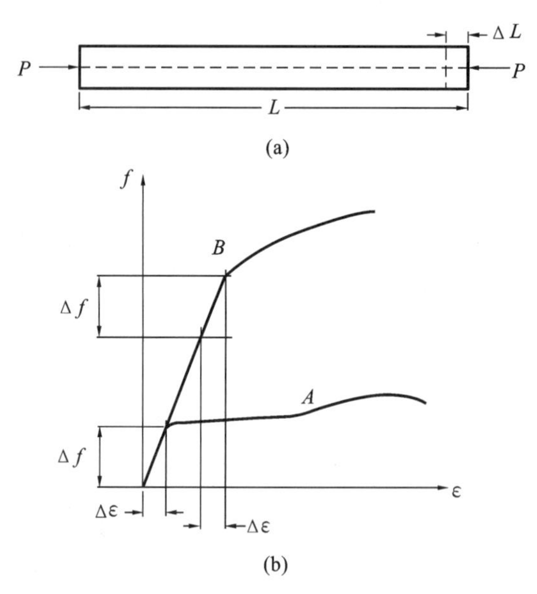

Fig. 16-1 Loss of prestress due to concrete shrinkage and creep

Alternatively, suppose that the beam is prestressed using high strength steel stressed to 150 000 psi. The elastic modulus of steel does not vary greatly, and the same value of 2.9×10^7 psi will be assumed here. Then in this case, the unit strain required to produce the desired stress in the steel is

$$\varepsilon_s = \frac{150\ 000}{2.9 \times 10^7} = 5.17 \times 10^{-3} \qquad 16\text{-}2$$

If shrinkage and creep strain are the same as before, the net strain in the steel after these losses is

$$\varepsilon_{s,net} = 5.17 \times 10^{-3} - 8.0 \times 10^{-4} = 4.37 \times 10^{-3} \qquad 16\text{-}3$$

and the corresponding stress after losses is

$$f_s = \varepsilon_{s,net} E_s = (4.37 \times 10^{-3}) \times (2.9 \times 10^7)\text{psi} = 127\ 000\ \text{psi} \qquad 16\text{-}4$$

This represents a stress loss of about 15 percent, compared with 100 percent loss in the beam using ordinary steel. It is apparent that the amount of stress lost because of shrinkage and creep is independent of the original stress in the steel. Therefore, the higher the original stress, the lower the percentage loss. This is illustrated graphically by the stress-strain curves of Fig. 16-1(b). Curve A is representative of ordinary reinforcing bars, with a yield stress of 60 000 psi, while curve B represents high tensile steel, with a tensile strength of 270 000 psi. The stress change Δf resulting from a certain change in strain $\Delta\varepsilon$ is seen to have much less effect when high steel stress levels are attained. Prestressing of concrete is therefore practical only when steels of very high strength are used.

Prestressing steel is most commonly used in the form of individual wires, stranded cable (strands) made up of seven wires (see Fig. 16-2), and alloy-steel bars. Virtually all strands in use are low-relaxation. ②

Fig. 16-2 Strands made up of seven wires

The tensile stress permitted by ACI code 18.5 in prestressing wires, strands, or bars is dependent upon the stage of loading. When the jacking force is first applied, a maximum stress of 0.80 f_{pu} or 0.94 f_{py} is allowed, whichever is smaller, where f_{pu} is the tensile strength of the steel and f_{py} is the yield strength. Immediately after transfer of prestress force to the concrete, the permissible stress is 0.74 f_{pu} or 0.82 f_{py}, whichever is smaller (except at post-tensioning anchorages where the stress is limited to 0.7 f_{pu}). The justification for a higher allowable stress during the stretching operation is that the steel stress is known quite precisely at this stage.[3] Hydraulic jacking pressure and total steel strain are quantities that are easily measured and quality control specifications require correlation of load and deflection at jacking. In addition, if an accidentally deficient tendon should break, it can be replaced; in effect, the tensioning operation is a performance test of the material.[4] The lower values of allowable stress apply after elastic shortening of the concrete, frictional loss, and anchorage slip have taken place. The steel stress is further reduced during the life of the member due to shrinkage and creep in the concrete and relaxation in the steel. ACI allowable stresses in prestressing steels are summarized in Table 16-1.

Table 16-1 Maximum permissible stresses in prestressing steel

1	Due to tendon jacking force but not greater than the lesser of $0.80f_{pu}$ and the maximum value recommended by the manufacturer of the prestressing steel or anchorage devices	$0.94\ f_{py}$
2	Immediately after prestress transfer but not greater than $0.74f_{pu}$	$0.82\ f_{py}$
3	Post-tensioning tendons, at anchorage devices and couplers, immediately after tendon anchorage	$0.70\ f_{pu}$

The strength and other characteristics of prestressing wires, strands, and bars vary somewhat between manufacturers, as do methods of grouping tendons and anchoring them.

Words and Expressions

rod [rɔd] *n.* 杆,棒,条
creep [kriːp] *n.* 徐变
relieve [ri'liːv] *v.* 减轻,解除
loss [lɔs] *n.* 损失
net [net] *adj.* 净的,纯的
stretch [stretʃ] *v.* 伸展,张拉
hydraulic [hai'drɔːlik] *adj.* 液压的
relaxation [ˌriːlæk'seiʃən] *n.* (预应力钢筋的)松弛
loss of prestress 预应力损失
high strength 高强度
alloy-steel 合金钢
frictional loss 摩擦损失
anchorage slip 锚具滑移

Notes

① Early attempts at prestressing concrete were unsuccessful because steel with ordinary structural strength was used. The low prestress obtainable in such rods was quickly lost due to shrinkage and creep in the concrete.

本句介绍了早期预应力混凝土的探索没有成功的原因,因为使用了普通结构强度的钢筋。而这种(非高强)钢筋所能得到的预应力值较低,进而因混凝土的收缩和徐变,而很快损失(殆尽)了。

② Prestressing steel is most commonly used in the form of individual wires, stranded cable (strands) made up of seven wires, and alloy-steel bars. Virtually all strands in use are low-relaxation.

本句介绍了最常见的预应力钢筋的(3 种)形式:单根钢丝、不超过 7 根钢丝扭结而成的钢绞线以及合金钢筋。(其中)几乎所有钢绞线都是低松弛的。

③ The justification for a higher allowable stress during the stretching operation is that the steel stress is known quite precisely at this stage.

本句意思为——当张拉阶段能很准确地知道钢材的应力时,张拉过程中可采用更高的允许应力值(否则,钢材中意料之外的高应力,可能会导致钢材断裂)。

④ In addition, if an accidentally deficient tendon should break, it can be replaced; in effect, the tensioning operation is a performance test of the material.

本句意思为——另外,如果一根偶然有缺陷的钢束断裂,还可以更换,事实上,张拉操作过程是对材料性能的一次检验(因为已经将预应力筋张拉至相当高的应力水平)。

Exercises

Ⅰ. Translate the following words into Chinese.

1. tensile strength ______________________________
2. yield strength ______________________________
3. relaxation in the steel ______________________________

Ⅱ. Translate the following words into English.

1. 长期应变______________________________
2. 混凝土收缩______________________________
3. 应力-应变曲线______________________________

Ⅲ. Translate the following sentences into Chinese.

1. Such changes in length of concrete have much less effect on prestress force if that force is obtained using highly stressed steel wires or cables.
2. It is apparent that the amount of stress lost because of shrinkage and creep is independent of the original stress in the steel.
3. Hydraulic jacking pressure and total steel strain are quantities that are easily measured and quality control specifications require correlation of load and deflection at jacking.

Ⅳ. Translate the following sentences into English.

1. 只有采用高强度钢材作为预应力筋,混凝土的预应力才能实现。
2. 在预应力混凝土构件的使用期间,由于混凝土的收缩、徐变以及预应力筋的松弛,预应力筋中的应力会继续降低。

Reading Materials

Concrete Prestressed Construction

Ordinarily, concrete of substantially higher compressive strength is used for prestressed structures than for those constructed of ordinary reinforced concrete. Most prestressed construction in the United States at present is designed for a compressive strength above 5000 psi. There are several reasons for this.

①High-strength concrete normally has a higher modulus of elasticity. This means a reduction in initial elastic strain under application of prestress force and a reduction in creep strain, which is approximately proportional to elastic strain. This results in a reduction in loss of prestress.

②In post-tensioned construction, high bearing stresses result at the ends of beams where the prestressing force is transferred from the *tendons*(预应力筋) to anchorage fittings, which bear directly against the concrete. This problem can be met by increasing the size of the anchorage fitting or by increasing the bearing capacity of the concrete by increasing its compressive strength. The latter is usually more economical.

③In pretension construction, where transfer by bond is customary, the use of high-strength concrete will permit the development of higher bond stresses.

④A substantial part of the prestressed construction in the United States is precast, with the concrete mixed, placed, and cured under carefully controlled conditions that facilitate obtaining higher strengths.

The strain characteristics of concrete under short-term and *sustained load*(持续荷载) assume an even greater importance in prestressed structures than in reinforced concrete structures because of the influence of strain on loss of prestress force. Strains due to stress, together with volume changes due to shrinkage and temperature changes, may have considerable influence on prestressed structures.

As for prestressing steels, the allowable stresses in the concrete, according to ACI code 18.4, depend upon the stage of loading and the behavior expected of the member. ACI code 18.3.3 defines three classifications of behavior, depending on the extreme fiber stress f_t at service load in the precompressed tensile zone. The three classifications are U, T, and C. Class U flexural members are assumed to behave as uncracked members. Class T members represent a transition between uncracked and cracked flexural members, while Class C members are assumed to behave as cracked flexural members. Permissible stresses for these three classifications are given in Table 16-2.

In Table 16-2, f'_{ci} is the compressive strength of the concrete at the time of initial prestress, and f'_{ci} the specified compressive strength of the concrete. In parts d and e of Table 16-2, sustained load is any part of the service load that will be sustained for a sufficient period of time to cause significant time-dependent deflections, whereas total load refers to the total service load, a part of which may be transient or temporary live load. Thus, sustained load would include dead load and may or may not include service live load, depending on its duration. If the live load duration is short or intermittent, the higher limit of part e is permitted.

Two-way slabs are designated as Class U flexural members*. Class C flexural members have no service level stress requirements but must satisfy strength and serviceability requirements. Service load stress calculations are computed based on uncracked section properties for Class U and T flexural members and on the cracked section properties for Class C members.

Table 16-2 Permissible stresses in concrete in prestressed flexural members

Condition†		Class		
		U	T	C
a	Extreme fiber stress in compression immediately after transfer	$0.60 f'_{ci}$	$0.60 f'_{ci}$	$0.60 f'_{ci}$
b	Extreme fiber stress in tension immediately after transfer (except as in c)†	$3\sqrt{f'_{ci}}$	$3\sqrt{f'_{ci}}$	$3\sqrt{f'_{ci}}$
c	Extreme fiber stress in tension immediately after transfer at the end of a simply supported member‡	$6\sqrt{f'_{ci}}$	$6\sqrt{f'_{ci}}$	$6\sqrt{f'_{ci}}$
d	Extreme fiber stress in compression due to prestress plus sustained load	$0.45 f'_{ci}$	$0.45 f'_{ci}$	—
e	Extreme fiber stress in compression due to prestress plus total load	$0.6 f'_{ci}$	$0.6 f'_{ci}$	—
f	Extreme fiber stress in tension f_t in precompressed tensile zone under service load	$\leq 7.5\sqrt{f'_{ci}}$	$>7.5\sqrt{f'_{ci}}$ and $\leq 12\sqrt{f'_{ci}}$	—

* There are no service stress requirements for Class C.

† Permissible stresses may be exceeded if it is shown by test or analysis that performance will not be impaired.

‡When computed tensile stresses exceed these values, bonded auxiliary prestressed or nonprestressed reinforcement shall be provided in the tensile zone to resist the total tensile force in the concrete computed with the assumption of an uncracked section.

科技英语写作技巧(1)——科技英语的基本特点(Ⅰ)

1. 总体特点。

科技英语是英语语言诸多变体中的一种。它与通常的文学语言表述有明显的不同。它的特点在于客观、直观、简练、准确。科技英语写作内容分类明确、特征突出,须有较好的英语基础,并在掌握了基本写作技巧后,才可以写出内容正确、文字流畅、表达准确的科技英语文章。

2. 语法特点。

(1)多用被动结构。

【Example】On the other hand, if the construction of buildings cannot *be efficiently organized*, it might result in personal or facilities injury or in destruction of equipment. Therefore, great care should *be taken* during the construction of buildings.

另一方面,如果房屋的建造不能被有效地组织,那就可能导致人身和设备的损伤,甚至设备的破坏,因此,应对房屋的建造过程给予足够的重视。

上例由2个句子构成,共出现了3个谓语动词,而其中运用了被动结构(斜体部分)的就有2个。被动结构在土木工程专业英语文献中的广泛应用由此可见一斑。

(2)多用非谓语动词结构。

非谓语动词形式是指分词、动词不定式和动名词,土木工程专业英语文献中常用非谓语动词形式有如下两个主要原因:

①非谓语动词形式能使语言结构紧凑,行文简练。

【Example】This was the first in a series of significant developments in the structural forms of concrete high-rise buildings, freeing them from the previous 20 to 25 story height limitations of the rigid-frame and flat plate systems.

例中的 freeing them from 分词短语,如果不用非谓语形式,只能用 which make them be free from 的从句形式,那样会使语句冗长,不符合土木工程专业文献的行文要求,即无法以较少的篇幅表达最多的重要信息。

该句是现在分词做定语的例子。除了现在分词,英语中的非谓语动词还有过去分词、动名词和动词不定式以及它们组成的短语。这些非谓语动词(或非谓语动词短语)可以在句子中担任主语、定语、状语、宾语、宾语补足语等,以相对简短的篇幅代替相应从句所表达的内容。

②非谓语动词形式能体现或区分出句中信息的重要程度。

【Example】High buildings using bracing systems are used to resist lateral loads.

上例中谓语动词 are used 表达了主要信息,现在分词短语 using 提供细节,即非重要信息。这是定式动词和非谓语动词在表达信息功能上的主要分工和区别。

(3)多用长句。

土木工程科学研究的目的是揭示土木工程设计及建造的规律并解释其特点及应用。这样的工作是一个复杂的程序,而且程序间的各个环节联系紧密,为了能准确、清晰地表达这种复杂现象及其之间紧密的关系，其专业英语文献需要用各种不同的主从复合句，而且会出现复句中从句套从句的现象。

One difficult in every building for heavy industry is the arrangement of the supports for an overhead crane, which is carried by a bridge across the building, and the bridge runs on rails, one along each wall.

One difficult…is…是该句子的主体结构，which 引导的非限制性定语从句修饰 an overhead crane，and 后的句子是对 bridge 的补充说明，one along each wall 又修饰 rails。

译文：重工业的所有厂房均存在一个难题，即如何布置桥式吊车的支撑结构，吊车装在横跨厂房的桥架上，桥架在沿两道纵墙布置的轨道上行驶。

(4)静态倾向显著。

英语倾向于用名词，因而叙述呈静态(static)；汉语倾向于用动词，因而叙述呈动态(dynamic)。S. Potter 在 *Changing English* 一书中曾指出英语"名词优于动词"的倾向。静态倾向在科技英语中表现得十分明显，主要表现在以下几个方面。

①形容词或动词的名词化。

形容词或动词的名词化主要指用名词来表达原来由动词或形容词所表达的概念，如用抽象名词来表达动作、行为、变化、状态、品质、情感等概念。名词化往往可以使表达比较简洁，结构紧凑。

例 1：Due to the *versatility* of the finite element method it can be applied to very complex problems.

译文：由于有限元法功能的多样性，它可以用于解决非常复杂的问题。

例 2：An important part of preliminary structural design is the *selection* of the structural system with *consideration* given to its relationship to construction economics.

译文：初步结构设计的一个重要部分是选择结构体系，同时应考虑它与建筑经济学的关系。

②名词连用。

名词连用是指在短语中，前面的名词作为形容词修饰最后一个名词。这种名词连用的短语结构简单，表达方便，词数少而信息量大，在现代英语里几乎俯拾皆是。

有限元分析：FEM analysis (finite element method analysis)

使用荷载：service load

设计规定：design specification

设计依据：design consideration

材料性能：material properties

格构梁：lattice girder

名词连用尽量略去虚词和其他次要的词，加强了英语的名词优势，也反映了英语追求简洁的趋势。这正是以叙事和推论为特点的科技英语这种文体所需要的。

另外，在名词和其他词的选择中，有些原则希望引起读者的注意。

①能用名词做定语的不用动名词做定语。

例如:measuring accuracy → measurement accuracy。

②可直接用名词或名词短语做定语的情况下,要少用 of 句型。

例如:accuracy of measurement → measurement accuracy。

③若有相应的形容词形式,要尽量用形容词做定语。

例如:experiment results → experimental results。

④下列情况下用动词的名词形式就不及直接用动词显得更简洁明快。

例如:Measurement of thickness of plastic sheet was made. → Thickness of plastic sheet was measured.

(5)科技英语更注重语法的地道、准确。

在科技英语表达中,能写出语法上正确的句子是最起码的要求。但在现实中,许多同学写出的句子往往是将脑子里的英语单词根据汉语的习惯罗列在一起,写出的句子“中国人不懂,外国人也不明白”,出现无主语、多谓语、主谓不一致等诸多现象。

究其原因,就是这些同学在进行英语写作或者翻译的时候没有认真“规划”英语句子。众所周知,汉语的表达习惯更注重词序,以及词与短语之间的内在逻辑关系,而英语常常使用各种连词、关系代词和关系副词来表示分句,以及主句与从句之间的各种关系。所以我们在撰写英语文章时一定要根据具体情况先分析句子成分,把“形式上无逻辑关系”的汉语句子成分“划归”成英语相应的句子成分,利用从句、非谓语动词,并适当“添加”相应的代词、连词和关系副词等,再适当巧妙运用如前所述的名词化等原则,写出地道准确的英语。

Lesson 17

LOSS OF PRESTRESS

The initial prestress force P_i immediately after transfer is less than the jacking force P_j because of elastic shortening of the concrete, slip at the anchorages, and frictional losses along the tendons. The force is reduced further, after a period of many months or even years, due to length changes resulting from shrinkage and creep of the concrete and relaxation of the highly stressed steel; eventually it attains its effective value P_e. Losses have no effect on the nominal strength of a member with bonded tendons, but overestimation or underestimation of losses may have a pronounced effect on service conditions including camber, deflection, and cracking.

The estimation of losses can be made on several different levels. Lump sum losses, used in the early development of prestressed concrete are now considered obsolete. Values of R, based on detailed calculations and verified in field applications are used in design offices, as are tables of individual loss contributions. For cases where greater accuracy is required, it is necessary to estimate the separate losses, taking account of the conditions of member geometry, material properties, and construction methods that apply. Accuracy of loss estimation can be improved still further by accounting for the interdependence of time-dependent losses, using the summation of losses in a sequence of discrete time steps. These methods will be discussed briefly in the following paragraphs.

Lump-sum estimates of losses

It was recognized very early in the development of prestressed concrete that there was a need for approximate expressions to be used to estimate prestress losses in design. Many thousands of successful prestressed structures have been built based on such estimates, and where member sizes, spans, materials, construction procedures, amount of prestress force, and environmental conditions are not out of the ordinary, so this approach is satisfactory. For such conditions, the American Association of State Highway and Transportation Officials (AASHTO) has recommended the values in Table 17-1 for preliminary design or for certain controlled precasting conditions. It should be noted that losses due to friction must be added to these values for post-tensioned members. These may be calculated separately by the equations of below.

The AASHTO recommended losses of Table 17-1 include losses due to elastic shortening, creep, shrinkage, and relaxation. Thus for comparison with R values for estimating losses, which included only the time-dependent losses due to shrinkage, creep, and relaxation, elastic shortening losses should be estimated by the methods discussed and deducted from the total.

Table 17-1 Estimate of prestress losses

Type of beam section	Level	Wires or strands with f_{pu}=235 250, or 270 ksi*
Rectangular beams, solid slabs	upper bound average	33.0 ksi 30.0 ksi
Box girder	upper bound average	25.0 ksi 23.0 ksi
I-girder	average	$33.0[1-0.15(f'_c-0.6)/6.0]+6.0$
Single T, double T hollow core and voided slab	upper bound average	$39.0[1-0.15(f'_c-0.6)/6.0]+6.0$ $39.0[1-0.15(f'_c-0.6)/6.0]+6.0$

* Values are for fully prestressed beams, reductions are allowed for partial prestress.

Losses due to friction are excluded. For low-relaxation strands, the values specified may be reduced by 4.0 ksi for box girders; 6.0 ksi for rectangular beams, solid slabs, and I girders; and 8.0 ksi for single T's, double T's, hollow core and voided slabs.

Estimate of separate losses

For most designs, a separate estimate of individual losses is made. Such an analysis is complicated by the interdependence of time-dependent losses. For example, the relaxation of stress in the tendons is affected by length changes due to creep of concrete. Rate of creep, in turn, is altered by change in tendon stress. ① In the following six subsections, losses are treated as if they occurred independently, although certain arbitrary adjustments are included to account for the interdependence of time-dependent losses. If greater refinement is necessary, a step-by-step approach may be used.

(1) Slip at the anchorages.

As the load is transferred to the anchorage device in post tensioned construction, a slight inward movement of the tendon will occur as the wedges seat themselves and as the anchorage itself deforms under stress. ② The amount of movement will vary greatly, depending on the type of anchorage and on construction techniques. Once this amount ΔL is determined, the stress loss is easily calculated from

$$\Delta f_{s,slip}=\frac{\Delta L}{L}E_s \tag{17-1}$$

It is significant to note that the amount of slip is nearly independent of the cable length. For this reason, the stress loss will be large for short tendons and relatively small for long tendons. The practical consequence of this is that it is most difficult to post-tension short tendons with any degree of accuracy.

(2)Elastic shortening of the concrete.

In pretensioned members, as the tendon force is transferred from the fixed abutments to the concrete beam, elastic instantaneous compressive strain will take place in the concrete, tending to reduce the stress in the bonded prestressing steel. The steel stress loss is

$$\Delta f_{s,\text{elastic}}=E_s\frac{f_c}{E_c}=nf_c \qquad 17\text{-}2$$

where f_c is the concrete stress at the level of the steel centroid immediately after prestress is applied:

$$f_c=-\frac{P_i}{A_c}\left(1+\frac{e^2}{r^2}\right)+\frac{M_0 e}{I_c} \qquad 17\text{-}3$$

If the tendons are placed with significantly different effective depths, the stress loss in each should be calculated separately.

In computing f_c by Eq. 17-3, the prestress force used should be that after the losses being calculated have occurred. It is usually adequate to estimate this as about 10 percent less than P_j.

In post-tensioned members, if all of the strands are tensioned at one time, there will be no loss due to elastic shortening, because this shortening will occur as the jacking force is applied and before the prestressing force is measured. [3] On the other hand, if various strands are tensioned sequentially, the stress loss in each strand will vary, being a maximum in the first strand tensioned and zero in the last strand. In most cases, it is sufficiently accurate to calculate the loss in the first strand and to apply one-half that value to all strands.

(3)Frictional losses.

Losses due to friction, as the tendon is stressed in post-tensioned members, are usually separated for convenience into two parts: curvature friction and wobble friction. The first is due to intentional bends in the tendon profile as specified and the second to the unintentional variation of the tendon from its intended profile. It is apparent that even a "straight" tendon duct will have some unintentional misalignment so that wobble friction must always be considered in post-tensioned work. The force at the jacking end of the tendon P_0, required to produce the force P_x at any point x along the tendon, can be found from the expression

$$P_0=P_x e^{Kl_x+\mu\alpha} \qquad 17\text{-}4(a)$$

Where: e—base of natural logarithms;

l_x—tendon length from jacking end to point x;

α—angular change of tendon from jacking end to point x, radians;

K—wobble friction coefficient, lb/lb per ft;

μ—curvature friction coefficient.

There has been much research on frictional losses in prestressed construction, particularly with regard to the values of K and μ. These vary appreciably, depending on construction methods and materials used. The values in Table 17-2, from ACI commentary R17.6, may be used as a guide.

Table 17-2 Friction coefficients for post-tensioned tendons

Type of tendon		Wobble coefficient K, per ft	Curvature coefficient μ
Grouted tendons in metal sheathing	Wire tendons	0.0010-0.0015	0.15-0.25
	High-strength bars	0.0001-0.0006	0.08-0.30
	Seven-wire strand	0.0005-0.0020	0.15-0.25
Unbonded tendons	Mastic-coated wire tendons	0.0010-0.0020	0.05-0.15
	Mastic-coated seven-wire strand	0.0010-0.0020	0.05-0.15
	Pregressed wire tendons	0.0003-0.0020	0.05-0.15
	Pregressed seven-wire strand	0.0003-0.0020	0.05-0.15

If one accepts the approximation that the normal pressure on the duct causing the frictional force results from the undiminished initial tension all the way around the curve, the following simplified expression for loss in tension is obtained:

$$P_0 = P_x(1 + Kl_x + \mu\alpha) \qquad 17\text{-}4(b)$$

where α is the angle between the tangents at the ends. The ACI code permits the use of the simplified form, if the value of $Kl_x + \mu\alpha$ is not greater than 0.3.

The loss of prestress for the entire tendon length can be computed by segments, with each segment assumed to consist of either a circular arc or a length of tangent.

(4) Creep of concrete.

Shortening of concrete under sustained load can be expressed in terms of the creep coefficient C_c. Creep shortening may be several times the initial elastic shortening, and it is evident that it will result in loss of prestress force. The stress loss can be calculated from

$$\Delta f_{s,creep} = C_c n f_c \qquad 17\text{-}5$$

Ultimate values of C_c for different concrete strengths for average conditions of humidity C_{cu} are given before.

In Eq. 17-5, the concrete stress f_c to be used is that at the level of the steel centroid, when the eccentric prestress force plus all sustained loads are acting. Eq. 17-5 can be used, except that the moment M_0 should be replaced by the moment due to all dead loads plus that due to any portion of the live load that may be considered sustained.

It should be noted that the prestress force causing creep is not constant but diminishes with the passage of time due to relaxation of the steel, shrinkage of the concrete, and length changes associated with creep itself. To account for this, it is recommended that the prestress force causing creep be assumed at 10 percent less than the initial value P_i. ④

(5) Shrinkage of concrete.

It is apparent that a decrease in the length of a member due to shrinkage of the concrete will be just as detrimental as length changes due to stress, creep, or other causes. The shrinkage strain ε_{sh} may vary between about 0.0004 and 0.0008. A typical value of 0.0006 may be used in lieu of specific data. The steel stress loss resulting from shrinkage is

$$\Delta f_{s,shrink} = \varepsilon_{sh} E_s \qquad 17\text{-}6$$

Only that part of the shrinkage that occurs after transfer of prestress force to the concrete needs to be considered. For pretensioned members, transfer commonly takes place just 17 hours after placing the concrete, and nearly all the shrinkage takes place after that time. However, post-tensioned members are seldom stressed at an age earlier than 7 days and often much later than that. About 15 percent of ultimate shrinkage may occur within 7 days, under typical conditions, and about 40 percent by the age of 28 days.

(6) Relaxation of steel.

The phenomenon of relaxation is similar to creep. Loss of stress due to relaxation will vary depending upon the stress in the steel. To allow for the gradual reduction of steel stress resulting from the combined effects of creep, shrinkage, and relaxation, the relaxation calculation can be based on a prestress force 10 percent less than P_i.

It is interesting to observe that the largest part of the relaxation loss occurs shortly after the steel is stretched. For stresses of 0.80 f_{pu} and higher, even a very short period of loading will produce substantial relaxation, and this in turn will reduce the relaxation that will occur later at a lower stress level. The relaxation rate can thus be artificially accelerated by temporary over tensioning. This technique is the basis for producing low-relaxation steel. ⑤

Loss estimation by the time-step method

The loss calculations of the preceding paragraphs recognized the interdependence of creep, shrinkage, and relaxation losses in an approximate way, by an arbitrary reduction of 10 percent of the initial prestress force P_i to obtain the force for which creep and relaxation losses were calculated. For cases requiring greater accuracy, losses can be calculated for discrete time steps over the period of interest. The prestress force causing losses during any time step is taken equal to the value at the end of the preceding time step, accounting for losses due to all causes up to that time. Accuracy can be improved to any desired degree by reducing the length and increasing the number of time steps.

A step-by-step method developed by the Committee on Prestress Losses of the Prestressed Concrete Institute uses only a small number of time steps and is adequate for ordinary cases.

Words and Expressions

camber	[ˈkæmbə]	*n.* 反弯度,曲面,弧
field	[fiːld]	*n.* 工地,现场
wobble	[ˈwɔbl]	*n.* 摆动,偏差
grout	[graut]	*v.* 灌(水泥)浆
unbonded	[ˌʌnˈbɔndid]	*adj.* 无黏结的
eccentric	[ikˈsentrik]	*adj.* 偏心的
detrimental	[ˌdetriˈmentəl]	*adj.* 有害的
frictional loss		摩擦损失
lump sum loss		总损失
over tensioning		超张拉
time-step method		时间步进法

Notes

① Such an analysis is complicated by the interdependence of time-dependent losses. For example, the relaxation of stress in the tendons is affected by length changes due to creep of concrete. Rate of creep, in turn, is altered by change in tendon stress.
这种分析的复杂性取决于时间的各种损失之间的相互关联。例如,钢束的应力松弛受到混凝土徐变引起的长度变化的影响。反之,徐变率又因钢束应力的变化而改变。

② As the load is transferred to the anchorage device in post tensioned construction, a slight inward movement of the tendon will occur as the wedges seat themselves and as the anchorage itself deforms under stress.
后张法结构中,当荷载传至锚具时,因楔块就位和应力下锚具本身的变形使钢束发生少量的回缩。(补充说明:进而导致钢束中的预拉应力减小。)

③ In post-tensioned members, if all of the strands are tensioned at one time, there will be no loss due to elastic shortening, because this shortening will occur as the jacking force is applied and before the prestressing force is measured.

后张法构件中，如果全部钢束都同时张拉，弹性缩短就不会引起损失。因为在千斤顶的力的作用下，预张力量测之前缩短已经出现。(补充说明：即达到指定张拉力时，混凝土构件的缩短在先前的张拉过程中已经完成。)

④ It should be noted that the prestress force causing creep is not constant but diminishes with the passage of time due to relaxation of the steel, shrinkage of the concrete, and length changes associated with creep itself. To account for this, it is recommended that the prestress force causing creep be assumed at 10 percent less than the initial value P_i.

应该注意，引起徐变的预张力并非常数，而是因为钢材的松弛、混凝土的收缩以及和徐变本身有关的长度变化等随时间进程而减小。考虑到这一点，建议将引起徐变的预张力假设为比初始值小 10% P_i。

⑤ It is interesting to observe that the largest part of the relaxation loss occurs shortly after the steel is stretched. For stresses of 0.80 f_{pu} and higher, even a very short period of loading will produce substantial relaxation, and this in turn will reduce the relaxation that will occur later at a lower stress level. The relaxation rate can thus be artificially accelerated by temporary over tensioning. This technique is the basis for producing low-relaxation steel.

有意思的是，经观察，大部分松弛损失在钢束张拉后很短时间内发生。当应力为 0.80 f_{pu}或更高时，即使在加载的很短时间内就出现相当大的松弛，随之又使发生在以后较低应力水平的松弛减小。因此，暂时的超张拉可人为地加速松弛率。有时这一技术可以用来减少损失。

Exercises

Ⅰ. Translate the following words into Chinese.

1. slip at the anchorage ______
2. relaxation of steel ______
3. relaxation rate ______

Ⅱ. Translate the following words into English.

1. 偏差摩擦系数______
2. 曲线摩擦系数______
3. 钢束形心______

Ⅲ. Translate the following sentences into Chinese.

1. For cases where greater accuracy is required, it is necessary to estimate the separate losses, taking account of the conditions of member geometry, mate-

rial properties, and construction methods that apply. Accuracy of loss estimation can be improved still further by accounting for the interdependence of time-dependent losses, using the summation of losses in a sequence of discrete time steps.

2. Losses due to friction, as the tendon is stressed in post-tensioned members, are usually separated for convenience into two parts: curvature friction and wobble friction.
3. For cases requiring greater accuracy, losses can be calculated for discrete time steps over the period of interest. The prestress force causing losses during any time step is taken equal to the value at the end of the preceding time step, accounting for losses due to all causes up to that time. Accuracy can be improved to any desired degree by reducing the length and increasing the number of time steps.

Ⅳ. Translate the following sentences into English.

1. 后张法构件中,如果各预应力筋相继张拉,每一钢束的应力损失将不同。
2. 后张法构件中,即使是一直线预应力筋的导管也会有些无意的偏差,因此后张法中总是必须考虑偏差系数。
3. 松弛引起的预应力损失,因预应力筋的应力大小而变化。

Reading Materials

Shape Section

One of the special features of prestressed concrete design is the freedom to select cross-section proportions and dimensions to suit the special requirements of the job at hand. The member depth can be changed, the web thickness modified, and the flange widths and thicknesses varied independently to produce a beam with nearly ideal proportions for a given case.

Several common precast shapes are shown in Fig. 17-1. Some of these are standardized and mass produced, employing reusable steel or *fiberglass*(玻璃纤维) forms. Others are individually proportioned for large and important works. The double T (see Fig. 17-1(a)) is probably the most widely used cross section in U. S. prestressed construction. A flat surface is provided, 4 to 12 ft wide. Slab thicknesses and web depths vary, depending upon requirements. Spans to 60 ft are not unusual. The single T (see Fig. 17-1(b)) is more appropriate for longer spans, to 120 ft, and heavier loads. The I and bulb T sections (see Fig. 17-1(c) and (d)) are widely used for bridge spans and roof girders up to about 140 ft, while the channel slab (see Fig. 17-1(e)) is suitable for floors in the intermediate span range. The box girder (see Fig. 17-1(f)) is advantageous for bridges of intermediate and major span. The inverted T section (see Fig. 17-1(g)) provides a bearing ledge to carry the ends of precast deck members spanning in the perpendicular direction. Local precasting plants can provide catalogs of available shapes. This information is also available in the *PCI Design Handbook*.

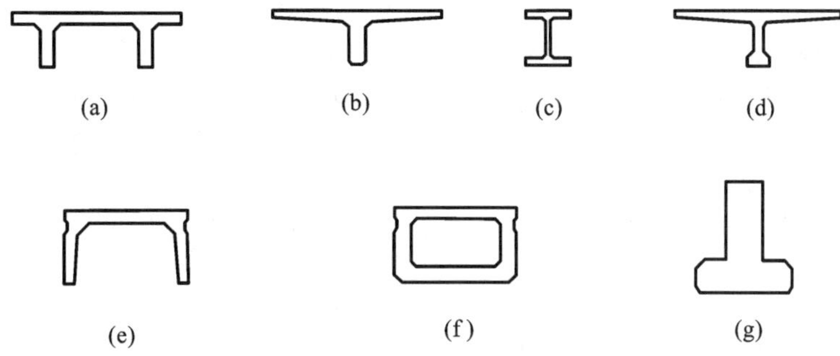

Fig. 17-1 Typical beam cross sections

(a) double T; (b) single T; (c) I girder; (d) bulb T;
(e) channel slab; (f) box girder; (g) inverted T

As indicated, the cross section may be symmetrical or unsymmetrical. An unsymmetrical section is a good choice: ①if the available stress ranges f_{1r} and f_{2r} at the top and bottom surfaces are not the same; ②if the beam must provide a flat, useful surface as well as offering load-carrying capacity; ③if the beam is to become a part of composite construction, with a cast-in-place slab acting together with a precast web; ④if the beam must provide support surfaces, such as shown in Fig. 17-1(g). In addition, T sections provide increased flexural strength, since the internal arm of the resisting couple at maximum design load is greater than for rectangular sections.

Generally speaking, I, T, and box sections with relatively thin webs and flanges are more efficient than members with thicker parts. However, several factors limit the gain in efficiency that may be obtained in this way. These include the instability of very thin overhanging compression parts, the vulnerability of thin parts to breakage in handling (in the case of precast construction), and the practical difficulty of placing concrete in very thin elements. The designer must also recognize the need to provide adequate spacing and concrete protection for tendons and anchorages, the importance of construction depth limitations, and the need for lateral stability if the beam is not *braced*(支撑) by other members against *buckling*(屈曲).

科技英语写作技巧(2)——科技英语的基本特点(Ⅱ)

3. 词汇特点。

1)多用语体正式的词汇。

英语中有很多意义灵活的动词短语,而土木工程专业英语文献中则多用与之对应的意义明确的单个词构成的动词,如用 absorb 代替 take in,用 discover 代替 find out。这类动词除了意义明确、精炼的特点外,还具有语体庄重、正式的特点。下面是一组对照,横线右边的词汇更适合于土木工程专业英语文献。

to use up—to exhaust
to push into—to insert
to increase in amount—to accumulate
to put in—to add
to pull towards—to attract
to use up—to consume
to think about—to consider
to find out—to discover
to take away—to remove
to pour out over the top—to overflow
to drive forward—to propel
to take up, to take in—to absorb
to speed up—to accelerate
to throw back—to reflect
to keep up—to maintain
to carry out—to perform
to breathe in—to inhale
to throw out, to get rid of—to eliminate
to push away—to repel
to get together—to concentrate
to fill up—to occupy
to pass on—to transmit,to transfer

同时,单个的英语词汇也有正式和非正式之分,土木工程专业英语文献属于正式文体,因此多用正式词汇。下面是一组对照,横线左边为非正式用语,右边为正式用语。

carry—bear
underwater—submarine
enough—sufficient
handbook—manual
try—attempt
get—obtain
about—approximately
finish—complete
hide—conceal
similar—identical
careful—cautious
feed—nourish
deep—profound
use—employ/utilize
oversee—supervise
buy—purchase
inner—interior
help—assist
stop—cease
leave—depart

2)科技英语词汇的构成。

随着科学技术的发展研究的深入,科技词汇的队伍也不断壮大,各个学科都建立了自己的专业词汇库。科技文章中除了普通词汇,还有半专业词汇、专业词汇等。

(1)半专业词汇。

半专业词汇的技术意义常常与其非技术意义不同,并且在不同的技术领域,半专业词汇的含义也不尽相同。所以,半专业词汇可以出现常用词汇专业化及同一词语词义多专业化的现象。例如:

concrete 具体→混凝土
reinforcement 加强→配筋,钢筋
detail 细节→构造,细部
stirrup 马蹬→箍筋
cap 帽子→承台,盖梁
frame 骨架→框架
beam 光束→梁
angle 角度→角钢
channel 槽道→槽钢
web 网→(工字钢等的)腹板,腹部
foot 脚→支座
coat 外套→涂装(动词)
aggregate 聚集→集料,骨料
bay 海湾→开间

joint 连接,关节→节点,(伸缩,沉降等)缝,(后浇)带
anchorage 停泊处→(预应力的)锚具
butt 顶撞,碰撞→对接(焊缝)
pile 堆,堆起→桩
support 支撑→支座
bearing 轴承→(桥梁中的)支座
leg 腿→(箍筋的)肢

(2)专业词汇。

专业词汇的语义精确,所指范围小。例如:cofferdam(围堰),tweeter(高音用扩音器),maglev(磁力悬浮火车)等。专业词汇采用了英语构词法,大量新词随着科技的发展不断产生。了解构词法在科技英语中的应用规律,对我们快速而准确理解词义具有重大作用。英语的基本词汇并不多,但有很大一部分构成型词汇,即通过构词法而形成的词汇。科技英语词汇常采用以下几种构词方法。

①复合词。

复合词是由两个或两个以上的旧词合成一个新词,例如复合名词、复合形容词、复合动词等。例如:

simply-supported 简支的
earth-moving machine 推土机
load-bearing capacity 承载力
high-rise building 高层建筑
dust-tight 防尘的
earth-fill 填土
pipeline 管线
water-proof 防水

②派生词。

派生词是指利用词的前缀与后缀作为词素构成新词。

前缀的特点:加前缀构成的新词只改变词义,不改变词类。

例:multistory 多层,superstructure 上部结构,substructure 下部结构,subsoil 地基土,antiseismic 抗震的,semidiameter 半径

后缀的特点:加后缀构成的新词可能改变也可能不改变词义,但一定改变词类。我们可以从一个词的后缀判别它的词类。如:

compression *n.* =compress+-ion(compress 是动词,-ion 是名词后缀)

liquidize *v.* =liquid+-ize(liquid 是名词,-ize 是动词后缀)

③缩略词。

缩略词是将较长的英语单词取其首部或者主干构成与原词同义的短单词,或者将组成词汇短语的各个单词的首字母拼接成一个大写字母的字符串。

例:Fig=figure　　lab=laboratory

RC=reinforced concrete(钢筋混凝土)
PC=prestressed concrete(预应力混凝土)
CAD=computer aided design(计算机辅助设计)
hyd. =hydrostatics(流体静力学)
KWIC=key-word-in-context(关键词在文内)
TS=tensile strength(抗拉强度)
FEM=finite element method(有限单元法)

Lesson 18

TYPES OF BRIDGES(1)

Any number of different methods may be used to classify bridges. Bridges can be classified according to materials (concrete, steel, or wood), usage (pedestrian, highway, or railroad), span (short, medium, or long), or structural form (slab, girder, truss, arch, suspension, or cable-stayed). None of these classifications are mutually exclusive. All seem to contain parts of one another within each other. For example, selection of a particular material does not limit the usage or dictate a particular structural form. On the other hand, unique site characteristics that require a long-span bridge with high vertical clearance limit the choices of materials and structural form.

The classification of bridge types in this presentation is according to the location of the main structural elements relative to the surface on which the user travels, that is, whether the main structure is below, above, or coincides with the deck line. ①

Main structure below the deck line

Arched and truss-arched bridges are included in this classification. Examples are the masonry arch, the concrete arch, the steel truss-arch, the steel deck truss, the rigid frame, and the inclined leg frame bridges. Striking illustrations of this bridge type are the Yangtze River RC arch bridge (see Fig. 18-1) in Wanxian, China and the Salginatobel Bridge in Switzerland.

Fig. 18-1 Wanxian RC arch bridge of Yangtze River, China

With the main structure below the deck line in the shape of an arch, gravity loads are transmitted to the supports primarily by axial compressive forces. At the supports, both vertical and horizontal reactions must be resisted. The arch rib can be solid or it can be a truss of various forms.

O'Connor (1971) summarizes the distinctive features of arch-type bridges as:

①The most suitable site for this form of structure is a valley, with the arch foundations located on dry rock slopes.

②The erection problem varies with the type of structure, being easiest for the cantilever arch and possibly most difficult for the tied arch.

③The arch is predominantly a compression structure. The classic arch form tends to favor concrete as a construction material.

④Aesthetically, the arch can be the most successful of all bridge types. It appears that through experience or familiarity, the average person regards the arch form as understandable and expressive. The curved shape is almost always pleasing.

Main structure above the deck line

Suspension, cable-stayed, and through-truss bridges are included in this category. Both suspension and cable-stayed bridges are tension structures whose cables are supported by towers. Examples are the Brooklyn Bridge and the East Huntington Bridge.

Suspension bridges (see Fig. 18-2) are constructed with two main cables from which the deck, usually a stiffened truss, is hung by secondary cables. Cable-stayed bridges have multiple cables that support the deck directly from the tower. Analysis of the cable forces in a suspension bridge must consider nonlinear geometry due to large deflections.

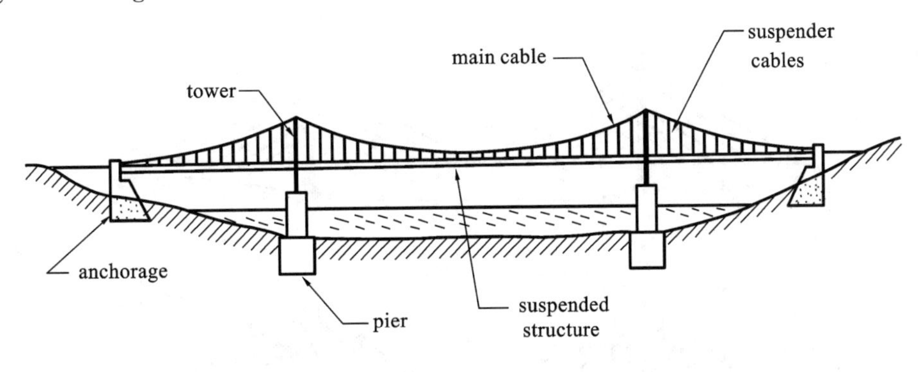

Elevation

Fig. 18-2 Typical suspension bridge

O'Connor (1971) gives the following distinctive features for suspension bridges.

①The major element of the stiffened suspension bridge is a flexible cable, shaped and supported in such a way that it can transfer the major loads to the towers and anchorages by direct tension. ②

②This cable is commonly constructed from high-strength wires either spun in situ or formed from component, spirally formed wire ropes. In either case the allowable stresses are high, typically of the order of 90 ksi (600 MPa) for parallel strands.

③The deck is hung from the cable by hangers constructed of high-strength wire ropes in tension.

④The main cable is stiffened either by a pair of stiffening trusses or by a system of girders at deck level.

⑤This stiffening system serves to control aerodynamic movements and limit local angle changes in the deck. It may be unnecessary in cases where the dead load is great.

⑥The complete structure can be erected without intermediate staging from the ground.

⑦It is the best alternative for spans over 2000 ft (600 m), and it is generally regarded as competitive for spans down to 1000 ft (300 m). However, even shorter spans have been built, including some very attractive pedestrian bridges.

Consider the following distinctive features for cable-stayed bridges (see Fig. 18-3) (O'Connor, 1971):

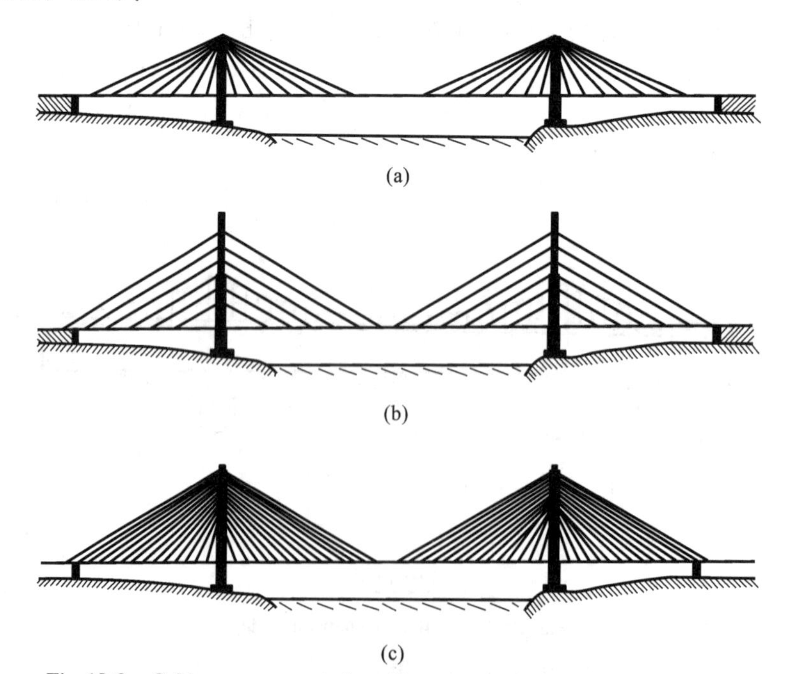

Fig. 18-3 Cable arrangements in cable-stayed bridges (Leonhardt, 1991)

(a) fan-shaped cable stayed bridge; (b) parallel-type cable bridge;
(c) modified fan-type cable stayed bridge

①As compared with the stiffened suspension bridge, the cables are straight rather than curved. As a result, the stiffness is greater. It will be recalled that the nonlinearity of the stiffened suspension bridge results from changes in the cable curvature and the corresponding change in bending moment taken by the dead-load cable tension. ③ This phenomenon can not occur in an arrangement with straight cables.

②The cables are anchored to the deck and cause compressive forces in the deck. For economical design, the deck system must participate in carrying these forces. In a concrete structure, this axial force compresses the deck.

③ Compared with the stiffened suspension bridge, the cable-braced girder bridge tends to be less efficient in supporting dead load but more efficient under live load. As a result, it is not likely to be economical on the longest spans. It is commonly claimed to be economical over the range 300-1100 ft (100-350 m).

④The cables may be arranged in a single plane, at the longitudinal centerline of the deck. This arrangement capitalizes on the torsion capacity inherent in a tubular girder system, and halves the number of shafts in the towers.

⑤The presence of the cables facilitates the erection of a cable-stayed girder bridge. Temporary backstays of this type have been common in the cantilever erection of girder bridges. Adjustment of the cables provides an effective control during erection.

Aerodynamic instability may be a problem with the stays in light rain and moderate winds. The water creates a small head (or bump) that disturbs the flow of wind around the cable. This disturbance creates an oscillatory force that may create large transverse movement of the stays. This phenomenon is sometimes called "dancing in the rain".

Words and Expressions

pedestrian	[pe'destriən]	*n.* 步行者,行人 *adj.* 徒步的,呆板的,乏味的
girder	['gə:də]	*n.* 大梁,纵梁
coincide	[ˌkəuin'said]	*v.* 一致,符合
deck	[dek]	*n.* 甲板,桥面板
aesthetically	[i:s'θetikəlli]	*adv.* 美学地,审美地,有审美感地
spun	[spʌn]	spin 的过去式和过去分词 *adj.* 纺成的
hanger	['hæŋə]	*n.* 吊索
aerodynamic	[ˌeərəudai'næmik]	*adj.* 空气动力学的
capitalize	[kə'pitəlaiz]	*v.* 以大写字母写,变成资本,利用

torsion	[ˈtɔːʃən]	*n*.[物][机] 扭转,转矩
halve	[hɑːv]	*v*. 二等分,平分,分享,减半
backstay	[ˈbækstei]	*n*.[船]桅杆,(斜拉桥的)背索
oscillatory	[ˈɔsileitəri]	*adj*. 摆动的
long-span bridge		大跨度桥
truss-arched bridge		桁架拱桥
masonry arch		砌石拱,砌体拱
inclined leg frame bridge		斜腿钢构桥
arch rib		拱肋
tied arch		系杆拱
through-truss bridge		下承式桁架桥
cable-stayed bridge		斜拉桥
suspension bridge		悬索桥
nonlinear geometry		几何非线性
stiffened suspension bridge		加劲悬索桥

Notes

① The classification of bridge types in this presentation is according to the location of the main structural elements relative to the surface on which the user travels, that is, whether the main structure is below, above, or coincides with the deck line.
这里介绍的桥的类型是根据主要结构单元相对于行驶的桥面的位置来进行划分的,也就是说,主要结构是位于桥面板的下方、上方还是与桥面板一体。

② The major element of the stiffened suspension bridge is a flexible cable, shaped and supported in such a way that it can transfer the major loads to the towers and anchorages by direct tension.
加劲悬索桥的主要单元是柔性索,其形状和支撑须便于以直接受拉的形式把主要荷载传递到主塔和锚锭。

③ It will be recalled that the nonlinearity of the stiffened suspension bridge results from changes in the cable curvature and the corresponding change in bending moment taken by the dead-load cable tension.
将可以看到,加劲悬索桥的非线性特性是由在静荷载作用下悬索曲率的变化和弯矩的相应变化产生的。

Exercises

Ⅰ. Translate the following words into Chinese.

1. medium-span bridge ________________________________

2. arch rib ____________________
3. cable-stayed bridge ____________________
4. peer review ____________________
5. masonry arch ____________________

Ⅱ. Translate the following words into English.

1. 拱肋____________________
2. 悬索桥____________________
3. 系杆拱桥____________________
4. 大跨度桥____________________
5. 下承式拱桥____________________
6. (斜拉桥的)背索____________________
7. 桥面板____________________

Ⅲ. Translate the following sentences into Chinese.

1. Arched and truss-arched bridges are included in this classification. Examples are the masonry arch, the concrete arch, the steel truss-arch, the steel deck truss, the rigid frame, and the inclined leg frame bridges.
2. Suspension bridges are constructed with two main cables from which the deck, usually a stiffened truss, is hung by secondary cables.
3. It is the best alternative for spans over 2000 ft (600 m), and it is generally regarded as competitive for spans down to 1000 ft (300 m). However, even shorter spans have been built, including some very attractive pedestrian bridges.

Ⅳ. Translate the following sentences into English.

1. 桥梁可以根据材料(混凝土、钢材和木材),用途(步行桥、公路桥和铁路桥),跨度(短、中等、大)或者结构形式(板桥、刚架桥、桁架桥、拱桥、悬臂桥或者斜拉桥)来进行分类。
2. 当桥面板下的主要结构为拱形时,重力荷载主要以轴向压力的形式传递给支撑结构。
3. 拉索锚固在桥面板上,从而在桥面板上产生压力。
4. 与加劲悬索桥相比,斜拉刚架桥能更有效地承载动荷载,而加劲悬索桥能更有效地承载静荷载。

Ⅴ. Answer the following questions briefly.

1. What is the cable commonly constructed from?
2. What is likely to be economical on the longest spans? why?
3. Which bridges are included in the category that main structure above the deck line? And how about the category that main structure below the deck line?

Reading Materials

Types of Bridges(2)

Main structure above the deck line (continued)

A truss bridge (see Fig. 18-4) consists of two main *planar trusses* (平面桁架) tied together with cross girders and *lateral bracing* (横撑) to form a three-dimensional truss that can resist a general system of loads. When the longitudinal stringers that support the deck slab are at the level of the *bottom chord* (下弦), this is a *through-truss bridge* (下承式桥) as shown in Fig. 18-5.

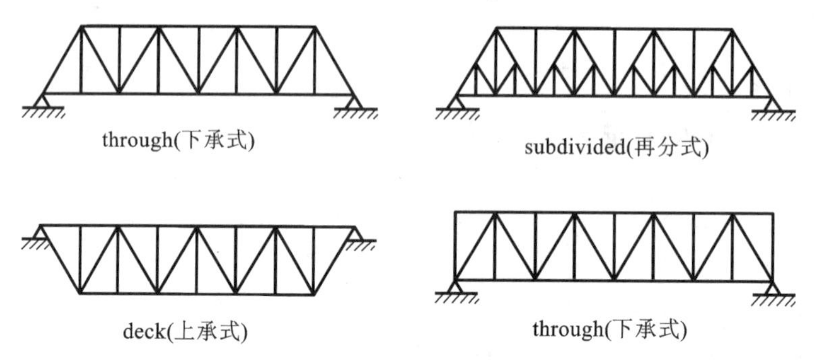

Fig. 18-4 Types of bridge trusses

Fig. 18-5 Greater New Orleans Through-Truss Bridge

O'Connor (1971) gives the following distinctive features for truss bridges:

①A bridge truss has two major structural advantages: the primary member forces are axial loads; the open web system permits the use of a greater overall depth than for an equivalent solid web girder. Both these factors lead to economy in material and a reduced dead weight. The increased depth also leads to reduced deflections, that is, a more rigid structure.

②The conventional truss bridge is most likely to be economical for medium spans. Traditionally, it has been used for spans intermediate between the plate girder and the stiffened suspension bridge. Modern construction techniques and materials have tended to increase the economical span of both steel and concrete girders. The cable-stayed girder bridge has become a competitor to the steel truss for the intermediate spans. These factors, all of which are related to the high *fabrication*(制作) cost of a truss, have tended to reduce the number of truss spans built in recent years.

③The truss has become almost the standard stiffening structure for the conventional suspension bridge, largely because of its acceptable aerodynamic behavior. Compared with alternative solutions, the *encroachment*(侵入) of a truss on the opening below is large if the deck is at the *upper chord*(上弦杆) level but is small if the traffic runs through the bridge, with the deck at the lower chord level. For railway overpasses carrying a railway above a road or another railway, the small construction depth of a through truss bridge is a major advantage. In some structures, it is desirable to combine both arrangements to provide a through truss over the main span with a small construction depth, and approaches with the deck at upper chord level.

Main structure coincides with the deck line

Girder bridges of all types are included in this category. Examples include slab (solid and voided), T-beam (cast-in-place), I-beam (precast or prestressed), wide-flange beam (composite and noncomposite), concrete box (cast-in-place and segmental, prestressed), steel box (orthotropic deck), and *steel plate girder*(钢板梁) (*straight and hunched*(加腋的)) bridges.

Illustrations of concrete slab, T-beam, prestressed girder, and box-girder bridges are shown in Fig. 18-6. A completed cast-in-place concrete slab girder-type bridges carry loads primarily in shear and flexural bending. This action is relatively inefficient when compared to axial compression in arches and to tensile forces in suspension structures. A girder must develop both compressive and tensile forces within its own depth. A *lever arm*(杠杆臂,内加臂) sufficient to provide the internal resisting moment separates these internal forces. Because the extreme fibers are the only portion of the cross section fully stressed, it is difficult to obtain an efficient distribution of material in a girder cross section. Additionally, stability concerns further limit the stresses and associated economy from a material utilization perspective. But from total economic perspective slab-girder bridges provide an economical and long-lasting solution for the vast majority of bridges. The U. S. construction industry is well tuned to provide this type of bridge. As a result, girder bridges are typical for short—to medium—span lengths, say <250 ft (75 m).

In highway bridges, the deck and girders usually act together to resist the applied load. Typical bridge cross sections for various types of girders are shown.

They include steel, concrete, and wood bridge girders with either cast-in-place or integral concrete decks. These are not the only combinations of gilders and decks but represent those covered by the approximate methods of analysis in the AASH-TO (2004) LRFD specifications.

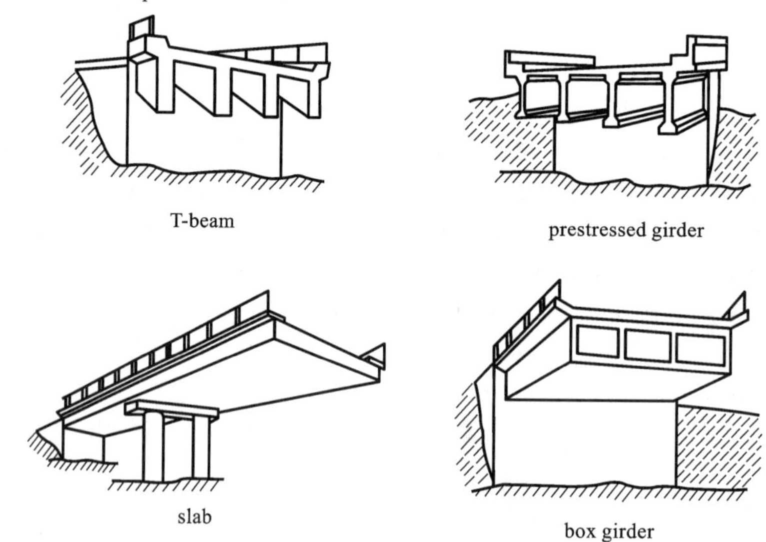

Fig. 18-6 Types of concrete bridges

Closing remarks on bridge types

For comparison purposes, typical ranges of span lengths for various bridge types are given in Table 18-1.

Table 18-1 Span lengths for various types of superstructure

Structural type	Material	Range of spans, ft(m)	Maximum span in service, ft(m)
slab	concrete	0-40(0-12)	
girder	concrete	40-1000(12-300)	988(301), Stolmasundent, Norway, 1998
	steel	100-1000(3-300)	984(300), Ponta Costa e Silva, Brazil, 1974
cable-stayed girder	steel	300-3500(90-1100)	3570(1088), Sutong, China, 2008
truss	steel	300-1800(90-550)	1800(550), Pont de Quebec, Canada, 1917(rail)
arch	concrete	300-1380(90-420)	1378(420), Wanxian, China, 1997
	steel truss	800-1800(240-550)	1800(550), Lupu, China, 2003
suspension	steel	1000-6600(300-2000)	6530(1991), Akashi Kaikyo, Japan, 1998

科技英语写作技巧(3)——数、量的表示法(Ⅰ)

1. 表示大小、尺度的方式。

(1)某物+have+a(n)+表示大小等的名词+of+数量。

➢ 这座山高2000米。

The mountain *has a height of* 2000 meters.

➢ 这些钢板厚2厘米。

These steel plates *have a thickness of* 2 centimeters.

➢ 这条马路宽10米。

This road *has a width of* 10 meters.

(2)某物+be+数量+表示大小的形容词(或in+表示大小的名词)。

前面提到的句子可以改写成如下形式而意思保持不变。

➢ The mountain is 2000 meters high.

OR: The mountain is 2000 meters in height.

➢ These steel plates are 2 centimeters thick.

OR: These steel plates are 2 centimeters in thickness.

➢ This road is 10 meters wide.

OR: This road is 10 meters in width.

2. 倍数、分数表示法。

(1)用倍数、分数原形表示的句型。

➢ 地球表面将近3/4为水所覆盖。

Nearly *three quarters of* the earth's surface is covered by water.

(2) {倍数 / 分数} + {the+名词 / that+后置定语(多为"of短语") / what从句}

➢ 这个梁的高度是那个的两倍。

This beam is 2 *times the height* of that one.

➢ 其速度为光速的十分之三。

Its speed is *three tenths that* of light.

3. 表示倍数增长的句型。

… *N* times+名词词组

… *N* times+as … as …

… *N* times+-er than …

…-er than … by *N* times

根据这一公式,汉语的倍数译成英语时要加一倍(指采用"… than …"句型时),读者要特别留意。例如:

➢ 这种普通的高铝水泥的价格大致比硅酸盐水泥贵两倍。

This common high-alumina cement costs roughly *three times more* than portland.

4. 用倍数表示"减少""降低"。

当 times, by a factor of 或-fold 与表示"减少""降低""小""少"这类词连用时,汉语习惯上用分数表达,一般不译成"减少……倍",而译作"减少……分之……"。

➢ Some steels can be drawn into wire five times finer than hair.

某些钢能被拉到头发丝的 1/5 那样细(一般不译成"……比头发丝细 5 倍")。

Lesson 19

SUBSTRUCTURE ELEMENTS OF BRIDGE

Of all the elements which make up a highway bridge, substructure elements present some of the most visibly striking features. A towering column or hammerhead pier can provide a certain sense of majesty which leaves a lasting image upon the traveler passing under a bridge. To be sure, for larger bridges, the superstructure elicits much, if not more, of the same feeling. For the majority of highway bridges, however, the only striking aspect of the design is found in the substructure. Even a row of reinforced earth modular units, snaking out along an abutment side slope, can be quite aesthetically pleasing.

If it can be said that the design of superstructure components varies greatly depending on geographic location and transportation department preferences, the same would be equally true for substructure components. In this section, we will cover the basic principles behind the design, rehabilitation, and maintenance of substructure components. Specifically the three major components which will be discussed: abutment, piers, and bearings.

Abutment

An abutment is a structure located at the end of a bridge which provides the basic functions of:

- Supporting the end of the first or last span;
- Retaining earth underneath and adjacent to the approach roadway, and, if necessary, supporting part of the approach roadway or approach slab.

To provide this functionality, a variety of abutment forms are used. The style of abutment chosen for a given bridge varies depending on the geometry of the site, size of the structure, and preferences of the owner. A simplification would be to think of an abutment as a retaining wall equipped with a bridge seat. The following discussion describes some of the most popular types of abutments in use, presents a design example for a typical abutment. ①

As mentioned above, most abutment types are variations on retaining wall configurations. With the exception of a crib wall, most any retaining wall system, when equipped with a bridge seat and designed to withstand the severe live loading conditions present in highway bridge structures, can be used as an abutment. ② Another difference between a conventional retaining wall system and a bridge abutment is that the latter is typically equipped with adjoining, flared walls known as wingwalls.

Wingwalls are designed to assist the principal retaining wall component of an abutment in confining the earth behind the abutment. The principal retaining wall component mentioned above is usually called the back wall or the stem of the abutment. The bridge seat, upon which the superstructure actually rests, is typically composed of either freestanding pedestals or a continuous breast wall. The pedestals or breastwall is designed to support bearings which in turn support individual primary member and transfer girder reactions to the foundation. They are located just in front of the backwall and sit on top of the abutment footing.

Piers

The development of bridge piers parallels the growth of the modern highway system. Previously, the use of bridge piers was confined to structures crossing rivers or railways. With the development of massive transportation networks, like the U. S. Interstate, the need for land piers to facilitate grade-separated highways increased dramatically.

A pier is a substructure which provides the basic function of supporting spans at intermediate points between end supports (abutments). Piers are predominately constructed using concrete, although steel and, to a lesser degree, timbers are also used. The concrete is generally conventionally reinforced. Prestressed concrete, however, is sometimes used as a pier material for special structures.

The basic design functions of a highway bridge pier can be summarized by the following list. In general, a pier is designed to:

- Carry its own weight;
- Sustain superstructure dead loads, live loads, and lateral loads;
- Transmit all loads to the foundation.

In addition to providing the structural functions detailed above, a properly designed pier should also be aesthetically pleasing and economize the use of materials as much as possible. Also, piers should be located so that they provide minimal interference with traffic passing underneath the structure.

Like abutments, piers come in a variety of configurations, shapes, and sizes. The type of pier selected will depend greatly on the form of superstructure present. The chosen single column piers for a prestressed concrete superstructure work well with the type of structure but would not be applicable for a composite steel structure composed of several primary members and a concrete deck.

Like an abutment, a pier has a bridge seat upon which the superstructure rests. In Fig. 19-1, this bridge seat consists of a hammerhead shaped pier cap on top of which are placed individual pedestals. The bearings are in turn placed on the top of the pedestals which rests the superstructure. It can be seen then, that in

Fig. 19-1 the pier is intended to support a superstructure composed of five primary members.

The bridge seat can be supported by a single column, multiple columns, a solid wall, or a group of piles. These supporting elements are in turn connected to the pier foundation which could be composed of footings, piles, or a combination thereof.

Fig. 19-1 A two column, concrete hammerhead pier under construction

Bearings

Bearings are mechanical systems which transmit loads from the superstructure to the substructure. In a way, bearings can be thought of as the interface between the superstructure and the substructure. In addition to transmitting vertical loads to the substructure component (i. e. , pier or abutment), a bearing also provides for movement due to thermal expansion and contraction as well as rotational movement associated with deflection of primary members.

The importance of bearings cannot be understated. Bearings which become frozen due to corrosion, clogged with debris, or fail to function as originally designed, can induce high stresses and potentially lead to failure of an individual span or an entire structure. The importance of bearings in a bridge design is often paid little (or no) attention because many bridge designers rely on manufacturers for the design of individual bearing units. In many cases, a bridge engineer will merely specify the loading conditions and movement that the bearing must accommodate and leave the detailed design of the individual units to the manufacturer of the bearing. This practice, however, is changing and bridge engineers are becoming increasingly aware of some of the nuances associated with design of bearings for highway bridges.

Like any other bridge component we have seen in the text thus far, bearings come in a variety of shapes and sizes. The applicability of certain types of bearings will vary depending on the loads and movement the bearing is required to sustain. Originally, steel mechanical bearing assemblies were the standard for most highway bridges. Within the past half century, however, the use of bearings made of a synthetic, elastomeric material has become increasingly popular. Indeed, the AASHTO specification for the design of elastomeric bearings did not first appear until 1961.

Words and Expressions

hammerhead	[ˈhæməhed]	*n.* 锤头
pier	[piə]	*n.* 桥墩,(凸式)码头
superstructure	[ˈsjuːpəˌstrʌktʃə(r)]	*n.* (建筑物、铁路、桥梁等的)上部结构,上层建筑
modular	[ˈmɔdjulə]	*adj.* [数]模的,模数的
geographic	[ˌdʒiəˈgræfik]	*adj.* 地理学的,地理的
rehabilitation	[ˈriː(h)əˌbiliˈteiʃən]	*n.* 复原,修复
stem	[stem]	*n.* 茎,干,词干,茎干 *v.* 起源于,阻止
pedestal	[ˈpedistəl]	*n.* 基架,底座,基础 *v.* 给……加底座,搁在台上
clog	[klɔg]	*n.* 木底鞋,障碍 *v.* 阻碍,阻塞
debris	[ˈdebriː]	*n.* 碎片,残骸
nuance	[njuːˈɑːns]	*n.* 细微差别
elastomeric	[iˌlæstəˈmerik]	*adj.* 弹性体的
towering column		塔柱
hammerhead pier		锤头状桥墩
retaining wall		挡土墙
bridge seat		桥支座
crib wall		框架式挡土墙
wingwall		翼墙,耳墙
backwall		后墙,背墙,挡土墙
breastwall		胸墙,防浪墙,挡土墙
grade-separated		等级划分
single column		独柱
multiple column		多柱
AASHTO		美国国家公路与运输协会标准

Notes

① The following discussion describes some of the most popular types of abutments in use, presents a design example for a typical abutment.
下面的讨论描述了一些最常用的桥台形式,介绍了一种典型桥台的设计实例。

② With the exception of a crib wall, most any retaining wall system, when equipped with a bridge seat and designed to withstand the severe live loading conditions present in highway bridge structures, can be used as an abutment.
除去一种框架挡土墙,大多数的挡土墙系统,当安装桥梁支座后可用来抵挡来自公路桥上巨大的活荷载作用时,可以当作一个桥台来使用。

Exercises

Ⅰ. Translate the following words into Chinese.

1. crib wall ______
2. retaining wall ______
3. bridge seat ______
4. grade-separated ______
5. multiple column ______

Ⅱ. Translate the following words into English.

1. 锤头状桥墩______
2. 高耸柱______
3. 独柱______
4. 耳墙______
5. 背墙______
6. 胸墙______

Ⅲ. Translate the following sentences into Chinese.

1. If it can be said that the design of superstructure components varies greatly depending on geographic location and transportation department preferences, the same would be equally true for substructure components.
2. Another difference between a conventional retaining wall system and a bridge abutment is that the latter is typically equipped with adjoining, flared walls known as wingwalls.
3. The development of bridge piers parallels the growth of the modern highway system.
4. In addition to providing the structural functions detailed above, a properly designed pier should also be aesthetically pleasing and economize the use of materials as much as possible.
5. The chosen single column piers for a prestressed concrete superstructure work well with the type of structure but would not be applicable for a composite steel structure composed of several primary members and a concrete deck.
6. The bearings are in turn placed on the top of the pedestals which rests the superstructure.

Ⅳ. Translate the following sentences into English.

1. 在所有组成公路桥的单元中，下部结构在视觉上呈现出最为显著的特征。
2. 为给定的桥梁选择桥墩的类型关键在于场地的几何形状、结构的规模和业主的设计要求。
3. 耳墙的设计是为了辅助桥墩的主挡土墙抵挡桥墩后面的泥土。

Ⅴ. Answer the following questions briefly.

1. What is the function of abutments?
2. What is the function of piers?
3. What is the function of bearings?
4. What is the function of earwall?

Reading Materials

Highway Bridge Loads and Load Distribution

Introduction

This article deals with highway bridge loads and load distribution as specified in the AASHTO load and resistance factor design (LRFD) specifications.

When proceeding from one component to another in bridge design, the controlling load and the controlling factored load combination will change. For example, permit vehicles, factored and combined for one load group, may control girder design for bending in one location. The standard design vehicular live load, factored and combined for a different load group, may control girder design for shear in another location. Still other loads, such as those due to seismic events, may control column and footing design.

Superstructure refers to the deck, beams or truss elements, and any other *appurtenances*(附属物) above the bridge soffit. Substructure refers to those components that support loads from the superstructure and transfer load to the ground, such as *bent caps*(盖梁,帽梁), columns, pier walls, footings, piles, pile extensions, and *caissons*(沉箱). Longitudinal refers to the axis parallel to the direction of traffic. Transverse refers to the axis perpendicular to the longitudinal axis.

Permanent loads

The LRFD specifications refer to the weights of the following as "permanent loads":

- The structure;
- Formwork which becomes part of the structure;
- Utility ducts or casings and contents;
- Signs;
- Concrete barriers;
- Wearing surface and/or potential deck overlay(s);
- Other elements deemed permanent loads by the design engineer and owner;
- Earth pressure, earth surcharge, and *downdrag* (下拉荷载).

The *permanent load* (永久荷载) is distributed to the girders by assigning to each all loads from superstructure elements within half the distance to the adjacent girder. This includes the *dead load*(静荷载) of the girder itself and the soffit, in the case of box girder structures. The dead loads due to concrete barrier, sidewalks and *curbs*(路缘), and sound walls, however, may be equally distributed to all girders.

Vehicular live loads

The AASHTO "design *vehicular live load*(汽车活荷载)", is a combination of a "*design truck*(设计车辆荷载)" or "*design tandem*(设计队列荷载)" and a "*design lane*(设计车道荷载)".

The AASHTO design truck is shown in Fig. 19-2. The variable axle spacing between the 145 kN loads is adjusted to create a critical condition for the design of each location in the structure. In the transverse direction, the design truck is 3 m wide and may be placed anywhere in the standard 3.6 m wide lane. The wheel load, however, may not be positioned any closer than 0.6 m from the lane line, or 0.3 m from the face of curb, barrier, or railing.

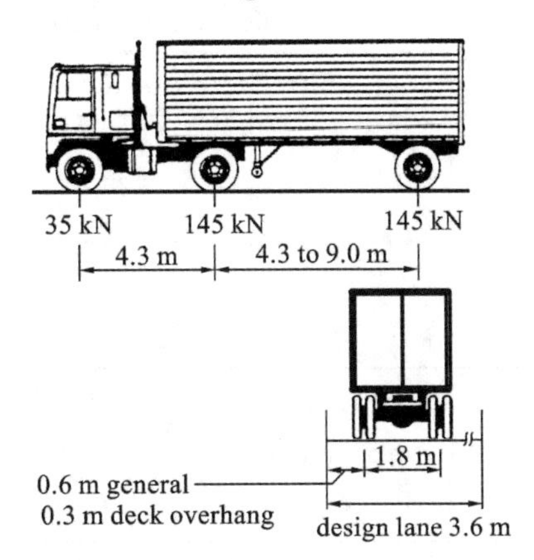

Fig. 19-2 AASHTO-LRFD design truck

The AASHTO design tandem consists of two 110 kN axles spaced at 1.2 m on center. The AASHTO design lane loading is equal to 9.3 N/mm and *emulates*(模拟) a caravan of trucks. Similar to the truck loading, the lane load is spread over a 3 m wide area in the standard 3.6 m lane. The lane loading is not interrupted except when creating an extreme force effect such as in "patch" loading of alternate spans. Only the axles contributing to the extreme being sought are loaded.

When checking an extreme reaction at an interior pier or *negative moment*(负力矩) between points of contra flexure in the superstructure, two design trucks with a 4.3 m spacing between the 145 kN axles are to be placed on the bridge with a minimum of 15 m between the rear axle of the first truck and the lead axle of the second truck. Only 90% of the truck and lane load is used. This procedure differs from the standard specification which used shear and moment riders.

Bridge substructure includes bent caps, columns, pier walls, *pile caps*(桩帽,

承台), spread footings, caissons, and piles. These components are designed by placing one or more design vehicular live loads on the traveled way as previously described for maximum reaction and negative bending moment, not exceeding the maximum number of vehicular lanes permitted on the bridge. This maximum may be determined by dividing the width of the traveled way by the standard lane width (3.6 m), and "rounding down", i.e., disregarding any fractional lanes. Note that: ①the traveled way need not be measured from the edge of deck if curbs or traffic barriers will restrict the traveled way for the life of the structure; ②the fractional number of lanes determined using the previously mentioned load distribution charts for girder design is not used for substructure design.

Multiple presence factors(多车道折减系数) modify the vehicular live loads for the probability that vehicular live loads occur together in a fully loaded state. The factors are shown in Table 19-1.

Table 19-1 Multiple presence factors

Number of loaded lanes	Multiple presence factors m
1	1.2
2	1.0
3	0.85
⩾3	0.65

These factors should be applied prior to analysis or design only when we use the lever rule or build three-dimensional model or work with substructures. Sidewalks greater than 600 mm can be treated as a fully loaded lane. If a two-dimensional girder line analysis is being done and distribution factors are being used for a beam-and-slab type of bridge, multiple presence factors are not used because the load distribution factors already consider three-dimensional effects. For the fatigue limit state, the multiple presence factors are also not used.

Pedestrian loads

Live loads also include pedestrians and bicycles. The LRFD specification calls for a 3.6×10^{-3} MPa load simultaneous with highway loads on sidewalks wider than 0.6 m. "Pedestrian- or bicycle-only" bridges are to be designed for 4.1×10^{-3} MPa. If the pedestrian- or bicycle-only bridge is required to carry maintenance or emergency vehicles, these vehicles are designed for omitting the dynamic load allowance. Loads due to these vehicles are infrequent and factoring up for dynamic loads is inappropriate.

Horizontal loads due to vehicular traffic and wind loads

Substructure design of vertical elements requires that horizontal effects of vehicular live loads be designed for. Centrifugal forces and braking effects are applied horizontally at a distance 1.80 m above the roadway surface. The LRFD specification provides wind loads as a function of base design wind velocity.

Exceptions to code-specified design loads

The designer is responsible not only for providing plans that accommodate design loads for the referenced design specifications, but also for any loads unique to the structure and bridge site. It is also the designer's responsibility to indicate all loading conditions designed for in the contract documents — preferably the construction plans. History seems to indicate that the next generation of bridge engineers will indeed be given the task of "improving" today's new structure. Therefore, the safety of future generations depends on today's designers doing a good job of documentation.

科技英语写作技巧(4)——数、量的表示法(Ⅱ)

5. 动词 increase，reduce。

本句型后面的数词为净增加数，意思是"增加(减少)几分之几"。

➢ 这个油田的原油产量在过去 5 年中平均增长 25%。

This oilfield has increased its annual output of crude oil *by an average of* 25 *percent* in the past five years.

➢ 由于引进了新设备，这家工厂的生产成本降低了三分之一。

The introduction of new equipment in this factory has reduced the cost of production (*by*) *one third*.

6. 百分数的表示。

用基数词(整数或小数)+percent(%)(有时分写成 per cent)。

➢ 水覆盖地球表面的约 70%的面积。

Water covers about *seventy percent of* the earth's surface.

➢ 按重量计的百分比

percentage by weight

➢ 按体积计的百分比

percentage by volume

➢ 这些量用百分比表示。

These quantities are expressed as *percentage*.

7. 用 double、treble 等表示。

… +double(treble/triple, quadruple)+ …

double 原意是"两倍、双倍"，可理解为：增加一倍、翻一番或是……的两倍。

treble/triple 原意是"三倍"，可理解为：增加两倍或是……的三倍。

quadruple 原意是"四倍"，可理解为：增加三倍或翻两番。

➢ 这个国家在过去十年中钢产量翻了一番。

This country has *doubled* her annual output of steel in the past ten years.

Lesson 20

SOIL PROPERTIES

Start digging with hand tools at almost any undeveloped site and the first layer of material exposed will be organic in nature—it will contain living organisms and their decomposing and decomposed arisings and will be mechanically weak. ① This topsoil is unsuitable for use for engineering purposes and is therefore set aside during construction, either for disposal or for reuse in planted areas upon completion of construction works.

Below the topsoil will lay an inorganic material whose properties may be reliably quantified and which is often suitable as a foundation for the pavement. This will sometimes be in the form of "rock"—such as limestone, granite, sandstone and the like—but is more often found to be a soil. A common definition of "rock" in this context is that the term shall include any hard natural or artificial material requiring the use of blasting or pneumatic tools for its removal but excluding small individual masses; "soil" is, by default, any other material found below ground level. Note that these descriptions differ from those used by the geologist, to whom all naturally occurring non-fluid materials are rock.

A further sub-division is often made in the case of soils—material may be either "suitable", that is to say it may be used as an engineering material somewhere in the works, perhaps as fill in an embankment, or it may be "unsuitable" in which case it is considered to be insufficiently stable for reuse elsewhere except for the forming of non-load bearing mounds. ② A typical application for unsuitable material is in the forming of mounds to act as visual and acoustic screening to a new road.

A naturally occurring soil will consist of particles of solid material, the voids between which are to a greater or lesser extent occupied by water. The physical properties of the soil will depend on the nature of these particles, the proportion that the volume of voids present bears to that of the whole body of material, and the amount of water present. In considering the suitability of a soil to support a pavement or other structure we are particularly concerned with the soil's ability to resist deformation caused by applied loads. Deformation may be caused by the soil tending to flow under the action of the applied load, or by its changing in volume.

Volume changes in soils resulting in settlement at the surface are caused by a limited rearrangement of the soil particles, resulting in a reduction in the proportion of voids present in the soil. Where the water or air originally present in the voids is able to leave the system quickly, as for example in the case of a coarse-grained soil with large voids between the particles, then such volume changes are not a great problem, since full settlement can be achieved by readily available techniques within a short period. In other cases, where the soil does not drain so freely, water in particular can only be expelled from the system by the continuous application of a substantial load over a long period. The first form of settlement, which may be induced by the use of compaction plant, is known as immediate settlement; the second, consolidation settlement.

Immediate settlement is a property of coarse-grained soils such as sands and gravels, while consolidation settlement is characteristic of soils consisting of very small individual particles such as clay and silt. For this and other reasons the engineer is often concerned with the range of particle sizes present in a soil.

The particle size distribution of a soil is often determined by sieve analysis. A full description of such a test is given in BS 1377 "Methods of testing soils for civil engineering purposes". The test apparatus consists of a series of sieves of gradually reducing standard sizes arranged vertically above one another with the coarsest at the top, all capable of being agitated by some mechanical means. The soil to be tested is dried and introduced to the top sieve, the whole stack of sieves shaken for some time, and the proportion retained on each sieve is determined by subtractive weighing and expressed as a proportion of the mass of the whole.

Knowledge of particle size distribution is of considerable value in assessing the likely behavior of the soil under a variety of circumstances. Soils with relatively large particles present free drainage paths for water in the soil and are therefore capable of being easily compacted; water can leave the system early. Very fine textured soils such as clays do not share this property and so are much less suitable for use as fill material, particularly where early stability is important such as beneath pavements and other structures.

Very fine grained soils exhibit the property that when slightly moist and when squeezed in the hand they may readily be formed into coherent lumps; the soils particles tend to stick together. This property is not shared by granular soils. Soils which stick together in this way are said to be cohesive; the most common cohesive is clay.

In their natural state clay particles are laminar, strongly bonded within themselves but only weakly bonded between one another. The laminar form of the particles tends to promote slip which is further assisted by the presence of water. Typically, clay contains water in two distinct ways—a water intimately linked to the clay particles by adsorption, and as free water. As the amount of water present in the system increases, so the plasticity—readiness to deform—of the clay increases;

wet clay is very plastic and is often unsuitable for use in or beneath the pavement.

The liquid limit of a soil is defined as the moisture content at which a soil passes from the plastic to the liquid state as defined by the liquid limit test. Liquid limit may be determined in various standard ways; these are described in BS 1377 and elsewhere. The most convenient and frequently used test is that using the cone penetrometer. The test consists of taking a sample of the soil passing a 425 micron sieve, mixing this thoroughly with water and placing it into a standard metal cup. A needle of standard shape and weight is then applied to the surface of the sample and is allowed to bear onto it for five seconds. The penetration into the sample is recorded to the nearest 0.1 mm. The moisture content of the soil sample is determined. This process is repeated several times with different moisture contents and a graph of penetration versus moisture content is drawn. The moisture content corresponding to a cone penetration of 20 mm is obtained from this graph and reported as a percentage to the nearest whole number as the liquid limit of the soil obtained by the cone penetrometer method. This method is preferred as it yields the most reproducible results.

Plastic limit is again a property of cohesive soils and is defined as the moisture content at which a soil becomes too dry to be a plastic condition as defined by the plastic limit test. The standard test consists of taking a 15 gram sample of the soil and mixing it with distilled water until it becomes plastic enough to be rolled into a ball. The material is then repeatedly rolled by hand into a thread of 3 mm diameter, the moisture content gradually being reduced by evaporation in the process. Eventually the soil becomes so dry that it crumbles when rolled into a 3 mm thread; this moisture content is reported as the plastic limit.

Thus both liquid and plastic limits are moisture contents, expressed as percentages by mass. Liquid limit of course always exceeds the magnitude of the plastic limit, and the difference between the two indicates the way in which the strength properties of the soil vary with changes in the moisture content. This numerical difference is known as the plasticity index (PI) thus:

$$\text{plasticity index} = \text{liquid limit} - \text{plastic limit}$$

and has been found to bear an empirical relationship to the CBR of soft cohesive soils. Since the liquid and plastic limits tests have been found to give superior repeatability to the CBR test, it is these former which are used in the pavement design process. We have seen that CBR depends at least in part on the service conditions—moisture content and surcharge.

The moisture content of a soil has an effect on its behavior in that very dry cohesive soils are difficult to compact, very wet granular or cohesive soils lack stability and are difficult to compact, and the moisture content of a soil can influence chemical reactions taking place within it. ③ Moisture content is defined as the ratio of the mass of water present in a body of soil to the mass of the dry soil particles. It is often expressed as a percentage and is measured by weighing a sample of the soil, drying it—usually in an oven at a temperature of 105 ℃, and weighing it again. The mass of water is determined by subtraction.

Words and Expressions

organism	[ˈɔːɡənizəm]	*n.* 有机物,有机体,生物
decompose	[ˌdiːkəmˈpəuz]	*v.* (使某物)变坏,腐烂
arisings	[əˈraiziŋz]	*n.* 副产品,废弃物,工业废料
pavement	[ˈpeivmənt]	*n.* 人行道,路面,铺面
limestone	[ˈlaimstəun]	*n.* 石灰岩
granite	[ˈɡrænit]	*n.* 花岗岩,花岗石
sandstone	[ˈsændstəun]	*n.* 砂岩
pneumatic	[njuːˈmætik]	*adj.* 由压缩空气操作(推动)的,风动的
embankment	[imˈbæŋkmənt]	*n.*(道路的)路堤,(河流的)岸堤
mound	[mɑund]	*n.* 土堆,土丘
expel	[ikˈspel]	*v.* 排出,驱逐,赶走,放逐
gravel	[ˈɡrævəl]	*n.* 砾石
clay	[klei]	*n.* 黏土
silt	[silt]	*n.* 粉土,淤泥
agitate	[ˈædʒiteit]	*v.* 搅动,摇动
drainage	[ˈdreinidʒ]	*n.* 排水,排水系统,废水,污水
texture	[ˈtekstʃə]	*n.* 质地,纹理
laminar	[ˈlæminə]	*adj.* 由薄片或层状体组成的,薄片状的
promote	[prəˈməut]	*v.* 提升,提拔,促进,推动,增进
micron	[ˈmaikrən]	*n.* 微米
needle	[ˈniːdl]	*n.* 针,编织针
penetration	[ˌpeniˈtreiʃən]	*n.* 穿过,渗透,突破,贯入
reproducible	[ˌriːprəˈdjuːsəbl]	*adj.* 能繁殖的,可重复的,可复写的
evaporation	[iˌvæpəˈreiʃən]	*n.* 蒸发(作用),汽化
crumble	[ˈkrʌmbl]	*v.* (把……)弄碎,(使)碎成细屑,碎裂
magnitude	[ˈmæɡnitjuːd]	*n.* 巨大,重要性,量值
surcharge	[ˈsəːtʃɑːdʒ]	*n.* 附加费
		v. 向……收取附加税

coarse-grained	粗骨粒的
immediate settlement	瞬时沉降
consolidation settlement	固结沉降
particle size	粒径

sieve analysis 筛分法
stack of sieves 套筒筛
liquid limit 液限
moisture content 含水量
cone penetrometer 锥式液限仪
whole number 整数
plastic limit 塑限
distilled water 蒸馏水
plasticity index 塑性指数
CBR=California bearing ratio 加州承载比

Notes

① Start digging with hand tools at almost any undeveloped site and the first layer of material exposed will be organic in nature—it will contain living organisms and their decomposing and decomposed arisings and will be mechanically weak.
当用手工工具开挖未曾开发利用的场地时,暴露出来的第一层材料将会是天然的有机物——土中含有生物有机物且它们正在分解或已经风化完成,土的力学性能较差。

② A further sub-division is often made in the case of soils—material may be either "suitable", that is to say it may be used as an engineering material somewhere in the works, perhaps as fill in an embankment, or it may be "unsuitable" in which case it is considered to be insufficiently stable for reuse elsewhere except for the forming of non-load bearing mounds.
对于土我们通常可以做进一步的划分——土既可以是适用材料,也就是说它可以在有些工程中做工程材料,如可用来填筑路堤;也可以是不可用材料,这种土没有足够的稳定性,不能在任何地方再使用,只能形成不能承受荷载的土堆。

③ The moisture content of a soil has an effect on its behavior in that very dry cohesive soils are difficult to compact, very wet granular or cohesive soils lack stability and are difficult to compact, and the moisture content of a soil can influence chemical reactions taking place within it.
土的含水量影响土的性质,因为非常干的黏性土很难被压实,而非常湿的粒料或黏性土缺乏稳定性也很难被压实,土的含水量也会影响土中化学反应的发生。

Exercises

Ⅰ. Translate the following words into Chinese.

1. immediate settlement ______
2. coarse-grained ______

3. stack of sieves ________________

4. moisture content ________________

5. cone penetrometer ________________

Ⅱ. **Translate the following words into English.**

1. 加州承载比________________

2. 液限________________

3. 筛分法________________

4. 固结沉降________________

5. 塑性指数________________

Ⅲ. **Translate the following sentences into Chinese.**

1. A common definition of "rock" in this context is that the term shall include any hard natural or artificial material requiring the use of blasting or pneumatic tools for its removal but excluding small individual masses.

2. The physical properties of the soil will depend on the nature of these particles, the proportion that the volume of voids present bears to that of the whole body of material, and the amount of water present.

3. Where the water or air originally present in the voids is able to leave the system quickly, as for example in the case of a coarse-grained soil with large voids between the particles, then such volume changes are not a great problem, since full settlement can be achieved by readily available techniques within a short period.

4. The test apparatus consists of a series of sieves of gradually reducing standard sizes arranged vertically above one another with the coarsest at the top, all capable of being agitated by some mechanical means.

5. The soil to be tested is dried and introduced to the top sieve, the whole stack of sieves shaken for some time, and the proportion retained on each sieve is determined by subtractive weighing and expressed as a proportion of the mass of the whole.

Ⅳ. **Translate the following sentences into English.**

1. 土的物理性质由土颗粒的性质、当前的空隙体积所占土体体积的比例和土体内现有的水量来决定。

2. 瞬时沉降是粗颗粒土的特性，比如砂土和砾石，而固结沉降是那些由细小的单颗粒组成的土的特征，如黏土和粉性土。

3. 自然状态的黏土颗粒是薄片状的，结构内部连接紧密，但彼此之间的连接很弱。

4. 塑限是黏性土的特性，其定义为土体从固态变化到可塑状态的界限含水量，通过塑限试验确定。

5. 液限总是大于塑限，两者的差值反映土体的强度随含水量的变化。

Ⅴ. Answer the following questions briefly.

1. What is of considerable value in assessing the likely behavior of the soil under a variety of circumstances?
2. Why are very fine textured soils much less suitable for use as fill material?
3. What is the moisture content of a soil? What is the liquid limit of a soil?

Reading Materials

Soil Mechanics

People use the land to live on, and build all sorts of structures: houses, roads, bridges, etcetera. It is the task of the geotechnical engineer to predict the behavior of the soil as a result of these human activities. The problems that arise are, for instance, the settlement of a road or a railway under the influence of its own weight and the traffic load, the margin of safety of an earth retaining structure (a dike, a quay wall or a *sheet pile wall*) (板桩墙), the earth pressure acting upon a tunnel or a *sluice*(水闸), or the allowable loads and the settlements of the foundation of a building. For all these problems soil mechanics should provide the basic knowledge.

Soil mechanics has become a distinct and separate branch of engineering mechanics because soils have a number of special properties, which distinguish from other materials. Its development has also been stimulated, of course, by the wide range of applications of soil engineering in civil engineering, as all structures require a sound foundation and should transfer its loads to the soil. The most important special properties of soils will be described briefly in this article.

Many engineering materials, not only metals, but also concrete and wood, exhibit linear stress-strain behavior, at least up to a certain stress level. This means that the deformations will be twice as large if the stresses are twice as large. This property is described by Hooke's law, and the materials are called linear elastic. Soils do not satisfy this law. For instance, in compression soil becomes gradually stiffer. At the surface sand will slip easily through the fingers, but under a certain compressive stress it gains an ever increasing stiffness and strength. This is mainly caused by the increase of the forces between the individual particles, which gives the structure of particles an increasing strength. This property is used in daily life by the packaging of coffee and other granular materials by a plastic envelope, and the application of vacuum inside the package. The package becomes very hard when the air is evacuated from it. In civil engineering the non-linear property is used to great advantage in the pile foundation for a building on very soft soil, underlain by a layer of sand. In the sand below a thick deposit of soft clay the stress level is high, due to the weight of the clay. This makes the sand very hard and strong, and it is possible to apply large compressive forces to the piles, provided that they are long enough to reach well into the sand.

In compression soils become gradually stiffer. In shear, however, soils become gradually softer, and if the shear stresses reach a certain level, with respect to the normal stresses, it is even possible that failure of the soil mass occurs. This means that the slope of a sand *heap*(堆), for instance in a *depot*(货栈,仓库) or in a dam, cannot be larger than about 30 or 40 degrees. The reason for this is that particles would slide over each other at greater slopes. As a consequence, many countries in deltas of large rivers are very flat. It has also caused the failure of dams and embankments all over the world, sometimes with very serious consequences for local population. Especially dangerous is that in very fine materials, such as clay, a steep slope is often possible for some time, due to capillary pressures in the water, but after some time these capillary pressures may vanish (perhaps because of rain), and the slope will fail.

A positive application of the failure of soils in shear is the construction of guard rails along highways. After a collision by a vehicle, the foundation of the guard rail will rotate in the soil due to the large shear stresses between this foundation and the soil body around it. This will dissipate large amounts of energy (into heat), creating a permanent deformation of the foundation of the rail, but the passengers, and the car, may be unharmed. Of course, the guard rail must be repaired after the collision, which can relatively easily be done with the aid of a heavy vehicle.

The deformations of a soil often depend upon time, even under a constant load. This is called creep. Clay and *peat*(泥煤, 泥炭块) exhibit this phenomenon. It causes structures founded on soft soils to show ever increasing settlements. A new road, built on a soft soil, will continue to settle for many years. For buildings such settlements are particular damaging when they are not uniform, as this may lead to cracks in the building.

The building of *dikes*(堤防) in the Netherlands, on compressible layers of clay and peat, results in settlements of these layers that continue for many decades. In order to maintain the level of the *crest*(顶部, 顶峰) of the dikes, they must be raised after a number of years. This results in increasing stresses in the subsoil, and therefore causes additional settlements. This process will continue forever. Before the construction of the dikes, the land was flooded now and then, with *sediment*(沉淀物, 沉积) being deposited on the land. This process has been stopped by man building dikes. Safety has an ever increasing price.

Sand and rock show practically no creep, except at very high stress levels. This may be relevant to the deformation of porous layers from which gas or oil are extracted.

A special characteristic of soil is that water may be present in the pores of the soil. This water contributes to the stress transfer in the soil. It may also be flowing with respect to the granular particles, which creates friction stresses between the fluid and the solid material. In many cases soil must be considered as a *two phase material*(二相材料). As it takes some time before water can be expelled from a soil mass, the presence of water usually prevents rapid volume changes.

It is also very important that lowering the water pressures in a soil, for instance by the production of *groundwater* (地下水) for drinking purposes, leads to an increase of the stresses between the particles, which results in settlements of the soil. This happens in many big cities, such as Venice and Bangkok, which may be threatened to be swallowed by the sea. It also occurs when a groundwater table is temporarily lowered for the construction of a dry excavation. Buildings in the vicinity of the excavation may be damaged by lowering the groundwater table. On a different scale the same phenomenon occurs in gas or oil fields, where the production of gas or oil leads to a volume decrease of the reservoir, and thus to subsidence of the soil. The production of natural gas from the large reservoir in Groningen is estimated to result in a subsidence of about 50 cm in the production time of the reservoir.

科技英语写作技巧(5)——“符合、一致”的表示法

常用动词及词组有：

accord 符合，一致　　　　　be in accord with 与……相一致
agree 符合，一致　　　　　be in agreement with 与……相一致
coincide 一致，相符　　　　be consistent with 与……相一致
conform(使)一致，(使)符合　be in conformity with 与……相一致
correspond to 符合于……
bring…into correspondence with 使……与……一致

常用名词有：

accord 一致，符合　　　　　agreement 符合，一致
coincidence 符合，一致　　conformity 符合，一致
consistency 一致(性)　　　correspondence 符合，一致

常用形容词有：

accordant 一致的　　　　　conformable 一致的，符合的
congruous 一致的　　　　　corresponding 一致的，符合的
consistent 一致的，符合的

➢ The experimental results agree/accord with the theoretical calculations.
实验结果与理论计算相符合。

➢ The theory does not correspond to the experimental facts.

这一理论与实验的实际情况不符。

➢ Our results coincide with those obtained by other workers.

我们的结果与其他工作者所取得的结果相一致。

科技英语写作技巧(6)——“原因、理由”的表示法

常用词语结构有：

because of/by virtue of /in view of/on account of+NP 由于……，因为……

due to/owing to+NP 由于……

for+NP(名词性短语) 由于……，因为……

from+NP 由于……

because/since/as… 因为……

in view of the fact that… 因为……，考虑到……这一事实

on account of the fact that… 因为……

owing to the fact that… 因为……

due to the fact that… 因为……，由于……(这一事实)

seeing that… 鉴于……，由于……(的缘故)

now that… 既然……，由于……

in that… 因为……，原因在于……

inasmuch as/insofar as 因为……，由于……

for+分句 因为……(不用于句首)

account for+NP/for the fact that… 是……的原因，说明……的原因

arise from+NP 由于……而引起

be attributed/ascribed to+NP 归因于……

result from+NP/from the fact that… 由于……所引起，由于……产生

The reason+why 分句+is+that 分句 (……之所以)……的原因是……

This is/explains+why 分句 这就是……的原因，这说明了为什么……

Lesson 21

FOUNDATION ENGINEERING INTRODUCTION

All engineering construction resting on the earth must be carried by some kind of interfacing element called a foundation. The foundation is the part of an engineering system that transmits to, and into, the underlying soil or rock the loads supported by the foundation and its self-weight. ① The resulting soil stresses—except at the ground surface—are in addition to those presently existing in the earth mass from its self-weight and geological history. ②

The term "superstructure" is commonly used to describe the engineered part of the system bringing load to the foundation, or substructure. The term "superstructure" has particular significance for buildings and bridges. However, foundations also may carry only machinery, support industrial equipment (pipes, towers, tanks), act as sign bases, and the like. For these reasons it is better to describe a foundation as that part of the engineering system that interfaces the load-carrying components to the ground.

It is evident on the basis of this definition that a foundation is the most important part of the engineering system.

The title foundation engineer is given to that person who by reason of training and experience is sufficiently versed in scientific principles and engineering judgment to design a foundation. We might say engineering judgment is a creative part of this design process.

Because of the heterogeneous nature of soil and rock masses, two foundations—even on adjacent construction sites—will seldom be the same except by coincidence. Since every foundation represents at least partly a venture into the unknown, it is of great value to have access to others' solutions obtained from conference presentations, journal papers, and textbook condensations of appropriate literature. The amalgamation of experience, study of what others have done in somewhat similar situations, and the site-specific geotechnical information to produce an economical, practical, and safe substructure design is application of engineering judgment.

The following steps are the minimum required for designing a foundation.

① Locate the site and the position of load. A rough estimate of the foundation load(s) is usually provided by the client or made in-house. Depending on the site or load system complexity, a literature survey may be started to see how others have approached similar problems.

② Physically inspect the site for any geological or other evidence that may indicate a potential design problem that will have to be taken into account when making the design or giving a design recommendation. Supplement this inspection with any previously obtained soil data.

③ Establish the field exploration program and, on the basis of discovery (or what is found in the initial phase), set up the necessary supplemental field test and any laboratory test program.

④ Determine the necessary soil design parameters based on integration of test data, scientific principles, and engineering judgment. Simple or complex computer analyses may be involved. For complex problems, compare the recommended data with published literature or engage another geotechnical consultant to give an outside perspective to the results.

⑤ Design the foundation using the soil parameters from step ④. The foundation should be economical and be able to be built by the available construction personnel. Take into account practical construction tolerances and local construction practices. Interact closely with all concerned (clients, engineers, architects, contractors) so that the substructure system is not excessively overdesigned and risk is kept within acceptable levels. [③] A computer may be used extensively (or not at all) in this step.

Geotechnical considerations will primarily be on strength and deformation and those soil water phenomena will affect strength and deformation. With the current trend to using sites with marginal soil parameters for major projects, methods to improve the strength and deformation characteristics through soil improvement methods will be briefly considered.

Foundations can be classified as shallow and deep foundations, depending on the depth of load-transfer from the structure to the ground. The definition of shallow foundations varies in different publications. BS 8004 (BSI, 1986) adopts an arbitrary embedment depth of 3 m as a way to define shallow foundations. In the context of this document, a shallow foundation is taken as one in which the depth to the bottom of the foundation is less than or equal to its least dimension.

The superstructure brings loads to the soil interface using column-type members. The load—carrying columns are usually of steel or concrete with allowable design compressive stresses on the order of 140^{+} MPa (steel) to 10^{+} MPa (concrete) and therefore are of relatively small cross-sectional area. The supporting capacity of the soil, from either strength or deformation considerations, is seldom over 1000 kPa but more often on the order of 200 to 250 kPa. This means the foundation is interfacing two materials with a strength ratio on the order of several hundred.

As a consequence the loads must be "spread" to the soil in a manner such that its limiting strength is not exceeded and resulting deformations are tolerable. Shallow foundations accomplish this by spreading the loads laterally, hence the term "spread footing". Where a spread footing (or simply footing) supports a single column, a mat is a special footing used to support several randomly spaced columns or to support several rows of parallel columns and may underlie a portion of or the entire building. The mat may also be supported, in turn, by piles or drilled piers. Foundations supporting machinery and such are sometimes termed bases. Machinery and the like can produce a substantial load intensity over a small area, so the base is used as a load-spreading device similar to the footing.

Deep foundations are analogous to spread footings but distribute the load vertically rather than horizontally. The terms drilled pier and drilled caisson are for the pile type member that is constructed by drilling 0.76^{+} m diameter hole in the soil, adding reinforce as necessary, and backfilling the cavity with concrete.

Rational design approaches require a greater geotechnical input including properly planned site investigations, field and laboratory testing, together with consideration of the method of construction. The use of rational methods to back-analyze results of loading tests on instrumented foundations or the monitored behavior of prototype structures has led to a better understanding of foundation behavior and enables more reliable and economical design to be employed.

Words and Expressions

foundation	[faun'deiʃən]	*n*. 基础
underlying	[ˌʌndə'laiiŋ]	*adj*. 下伏的,基础的,潜在的
soil	[sɔil]	*n*. 泥土,土壤,土地
rock	[rɔk]	*n*. 岩石
heterogeneous	[ˌhetərəu'dʒiːniəs]	*adj*. 多种多样的,各向异性的
amalgamation	[əˌmælgə'meiʃən]	*n*. 融合,合并
exploration	[ˌeksplə'reiʃən]	*n*. 勘察,探索,勘探
geological history		地质历史
foundation engineer		基础工程师
laboratory test		室内试验
soil improvement		地基加固
deep foundation		深基础
shallow foundation		浅基础
spread footing		扩展基础
drilled pier		钻孔墩
drilled caisson		钻孔沉井

Notes

① The foundation is the part of an engineering system that transmits to, and into, the underlying soil or rock the loads supported by the foundation and its self-weight. 基础为整个工程体系的一部分,它用于将其所承受的上部荷载及其自重传递至其下的土层或岩层。

② The resulting soil stresses — except at the ground surface — are in addition to those presently existing in the earth mass from its self-weight and geological history.
这些荷载导致除地表土之外的地基土体中产生附加于原有应力之上的应力,而原有应力是由土体自重及地质历史所形成的。

③ Take into account practical construction tolerances and local construction practices. Interact closely with all concerned (clients, engineers, architects, contractors) so that the substructure system is not excessively overdesigned and risk is kept within acceptable levels.
应考虑到实际的施工容许偏差以及当地的施工实践,还应与相关方面(业主、工程师、建筑师、承包方)充分交换意见,使得地下结构体系既可避免过度设计,且风险也被控制在容许的范围之内。

Exercises

Ⅰ. Translate the following words into Chinese.

1. underlying soil ____________________
2. heterogeneous nature ____________________
3. the site-specific geotechnical information ____________________
4. field exploration program ____________________
5. soil improvement methods ____________________

Ⅱ. Translate the following words into English.

1. 地质历史____________________
2. 沉井____________________
3. 扩展基础____________________
4. 地基加固____________________
5. 筏形基础____________________

Ⅲ. Translate the following sentences into Chinese.

1. The term "superstructure" is commonly used to describe the engineered part of the system bringing load to the foundation, or substructure.
2. For these reasons it is better to describe a foundation as that part of the engineering system that interfaces the load-carrying components to the ground.
3. Because of the heterogeneous nature of soil and rock masses, two founda-

tions—even on adjacent construction sites—will seldom be the same except by coincidence.

4. Geotechnical considerations will primarily be on strength and deformation and those soil water phenomena will affect strength and deformation.
5. The terms drilled pier and drilled caisson are for the pile type member that is constructed by drilling 0.76^{+} m diameter hole in the soil, adding reinforce as necessary, and backfilling the cavity with concrete.

Ⅳ. Translate the following sentences into English.

1. 必须在某种意义上将上部结构的荷载扩散到地基土体中,这样才会使土体中的应力不超过土体的极限强度。
2. 所有建造在地球上的工程结构物,都必须支承在一种称为“基础”的结构构件上。
3. 制定野外勘探计划,并根据最初阶段发现的情况提出必须补充的现场试验以及室内试验计划。
4. 深基础与扩展基础相似,所不同的是向竖直方向而不是向水平方向扩散荷载。
5. 根据试验数据、科学原理和工程判断的综合,确定必需的土体设计参数。

Ⅴ. Answer the following questions briefly.

1. What is the foundation? Try to state it in your own words.
2. What are the steps required for designing a foundation?
3. What is the difference between shallow foundation and deep foundation?

Reading Materials

General Requirements and Additional Considerations of Foundations

General requirements

Foundation elements must be proportioned both to interface with the soil at a safe stress level and to limit settlements to an acceptable amount. In the past 50 years few buildings (but numerous embankment types) have failed as a result of overstressing the underlying soil. However, excessive settlement problems are fairly common and somewhat concealed since only the most spectacular ones get published.

The variability of soil in combination with unanticipated loads or subsequent soil movements (e.g. earthquakes) can result in settlement problems over which the designer may have little control. In other words, current state-of-the-art design methods may greatly reduce the likelihood (risk factor) of settlement problems but do not generally provide a risk-free project.

A major factor that greatly complicates foundation design is that the soil parameters used for the design are obtained before the project is started. Later when the foundation is in place, it is on (or in) soil with properties that may be considerably modified from the original, either from the construction process or from installing the foundation. That is, the soil may be excavated and/or replaced and compacted; *excavations*(开挖) tend to remove load and allow the underlying soil to expand; *pile driving* (打桩) usually makes soil more dense, etc. Any of these events either directly alters the soil (replacement) or modifies the initially estimated soil strength parameters.

Another factor that encourages conservative design is the fact that many geotechnical engineers tend to imply that their talents (and design recommendations) are better than those of the competition. This generates a false sense on the part of the client that using that geotechnical engineer will produce a minimum cost foundation. When this happens and problems later occur (unanticipated soil strata, water, excessive settlement, or whatever), the client is very likely to litigate (i. e. sue). This possibility means that geotechnical engineers should be candid about the status of the state of the art in this specialty and make the client fully aware that precise soil parameters are difficult if not impossible to quantify and that at least some design conservatism is prudent.

Design conservatism means that any two design firms are unlikely to come up with exactly the same soil parameters and final foundation design. It would not be unusual for one firm to recommend the base contact pressure q_0 to be, say, 200 kPa whereas another might recommend 225 or even 250 kPa—both with the use of *spread footings*(扩展基础). There might be a problem in ethics, however, if one firm recommended 200 kPa and the other recommended only 100 kPa, which would require a *mat foundation*(筏板基础) or the use of piles. One of the recommendations is either overly optimistic (200 kPa) or realistic; the other is either realistic or overly conservative. Being excessively conservative is an ethics problem, unless the client is aware of the several alternatives and accepts the more conservative recommendation as being in his or her best interests.

In summary, a proper design requires the following.

① Determining the building purpose, probable service-life loading, type of framing, soil profile, construction methods, and construction costs.

② Determining the client's/owner's needs.

③ Making the design, but ensuring that it does not excessively degrade the environment, and provides a margin of safety that produces a tolerable risk level to all parties: the public, the owner, and the engineer.

Additional considerations

The previous section outlined in general terms requirements shall be met in designing a foundation in terms of settlement and soil strength. We will now outline a number of additional considerations that have to be taken into account at specific sites.

① Depth must be adequate to avoid lateral squeezing of material from beneath the foundation for footings and mats. Similarly, excavation for the foundation must take into account that this can happen to existing building footings on adjacent sites and requires suitable precautions be taken. The number of settlement cracks that are found by owners of existing buildings when excavations for adjacent structures begin is truly amazing.

② Depth of foundation must be below the zone of seasonal volume changes caused by freezing, thawing, and plant growth. Most local building codes will contain minimum depth requirements.

③ The foundation scheme have to consider expansive soil conditions. Here the building tends to capture upward-migrating soil water vapor, which condenses and saturates the soil in the interior zone, even as normal perimeter evaporation takes place. The soil in a distressingly large number of geographic areas tends to swell in the presence of substantial moisture and carry the foundation up with it.

④ In addition to compressive strength considerations, the foundation system must be safe against *overturning*(倾覆), sliding, and any uplift (flotation).

⑤ System must be protected against corrosion or deterioration due to harmful materials present in the soil. Safety is a particular concern in reclaiming sanitary landfills but has application for marine and other situations where chemical agents that are present can corrode metal pilings, destroy wood sheeting/piling, cause adverse reactions with *Portland cement*(普通水泥,硅酸盐水泥) in concrete footings or piles, and so forth.

⑥ Foundation system should be adequate to sustain some later changes in site or construction geometry and be easily modified when changes in the superstructure and loading become necessary.

⑦ The foundation should be buildable with available construction personnel. For one-of-a-kind projects there may be no previous experience. In this case, it is necessary that all concerned parties carefully work together to achieve the desired result.

⑧ The foundation and site development must meet local environmental standards, including determining if the building is or has the potential for being contaminated with hazardous materials from ground contact, for example, *radon*(氡) or *methane gas*(甲烷). Adequate air circulation and ventilation within the building are the responsibility of the mechanical engineering group of the design team.

Although not all of the preceding are applicable to a given project, it is readily apparent that those are tend to introduce much additional uncertainty into the system. This makes the application of engineering judgment an even more important ingredient in the design process.

科技英语写作技巧(7)——表示“原因、结果”的常用词语

表示原因、结果的例句：

- A mixing of all wavelengths causes/results in/produces/induces a white light.
- White light is caused by/due to/induced by/a result of/produced by a mixing of wavelengths.
- If/When/As all the wavelengths are mixed, a white light is produced.
- A white light is produced if/when/as all the wavelengths are mixed.

表达原因和结果的常用词语：

so, thus, consequently, as a result, hence, therefore, accordingly, for this reason, owing to, since, because of, the result of, result in, the effect of, the consequence of, have an effect on, the reason for, the cause of, it follows that…, now that, seeing that, for fear that, such…that, so as…to, make…possible, make it possible/impossible for…to…, due to, as, so that, so…that, thanks to, out of, owe…to, etc.

Lesson 22

PILES

Piles are structural members of timber, concrete, and/or steel that are used to transmit surface loads to lower levels in the soil mass. This transfer may be by vertical distribution of the load along the pile shaft or a direct application of load to a lower stratum through the pile point. A vertical distribution of the load is made using a friction (or floating) pile and a direct load application is made by a point, or end-bearing, pile. This distinction is purely one of convenience since all piles carry load as a combination of side resistance and point bearing except when the pile penetrates an extremely soft soil to a solid base.

Piles are commonly used (refer to Fig. 22-1) for the following purposes.

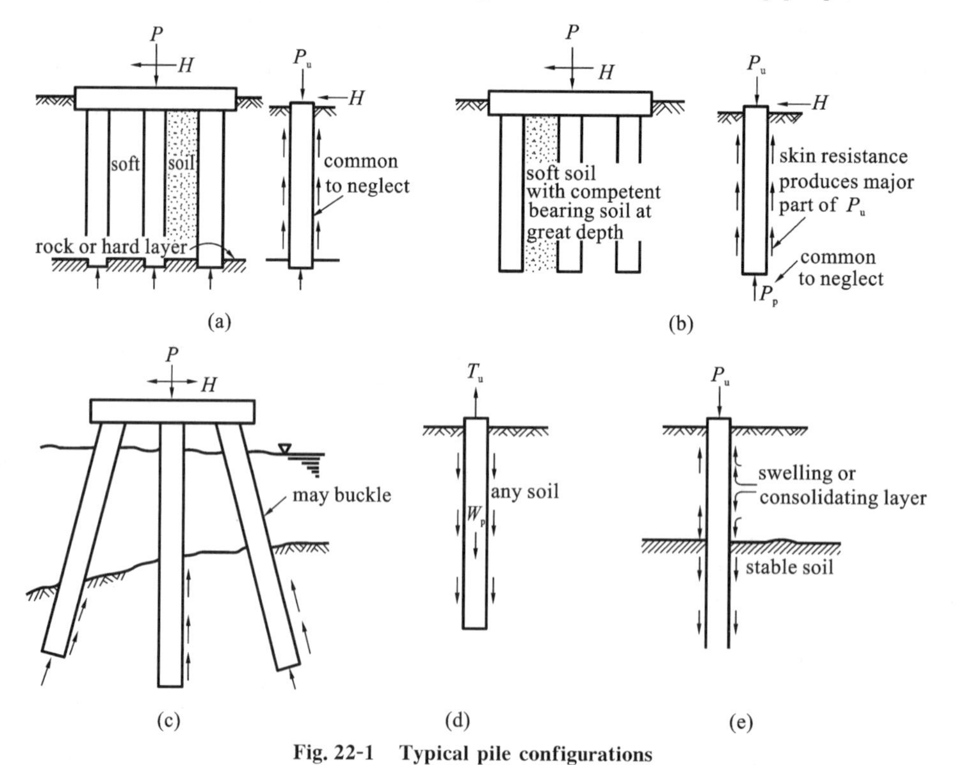

Fig. 22-1 Typical pile configurations

(a) group and single pile on rock or very firm soil stratum; (b) group or single pile "floating" in soil mass; (c) offshore pile group; (d) tension pile; (e) pile penetrating below a soil layer that swells (shown) or consolidates

① To carry the superstructure loads into or through a soil stratum. Both vertical and lateral loads may be involved.

② To resist uplift, or overturning, forces, such as for basement mats below the water table or to support tower legs subjected to overturning from lateral loads such as wind.

③ To compact loose, cohesionless deposits through a combination of pile volume displacement and driving vibrations. These piles may be later pulled.

④ To control settlement when spread footing or a mat is on a marginal soil or is underlain by a highly compressible stratum.

⑤ To stiffen the soil beneath machine foundations to control both amplitudes of vibration and the natural frequency of the system.

⑥ As an additional safety factor beneath bridge abutments and/or piers, particularly if scour is a potential problem.

⑦ In offshore construction to transmit loads above the water surface through the water and into the underlying soil. This case is one in which partially embedded piling is subjected to vertical (and buckling) as well as lateral loads.

Piles are sometimes used to control earth movements (for example, landslides). The reader should note that power poles and many outdoor sign poles may be considered as partially embedded piles subject to lateral loads. Vertical loads may not be significant, although buckling failure may require investigation for very tall members.

A pile foundation is much more expensive than spread footings and likely to be more expensive than a mat. In any case great care should be exercised in determining the soil properties at the site for the depth of possible interest so that one can as accurately as possible determine whether a pile foundation is needed and, if so, that neither an excessive number nor lengths are specified.[①] A cost analysis should be made to determine whether a mat or piles, in particular the type (steel, concrete, etc.), are more economical. In those cases where piles are used to control settlement at marginal soil sites, care should be taken to utilize both the existing ground and the piles in parallel so that a minimum number are required.

Piles are inserted into the soil via a number of methods.

① Driving with a steady succession of blows on the top of the pile using a pile hammer. This produces both considerable noise and local vibrations, which may be disallowed by local codes or environmental agencies and, of course, may damage adjacent property.

② Driving using a vibratory device attached to the top of the pile. This method is usually relatively quiet, and driving vibrations may not be excessive. The method is more applicable in deposits with little cohesion.

③ Jacking the pile. This technique is more applicable for short stiff members.

④ Drilling a hole and either inserting a pile into it, or more commonly, filling the cavity with concrete, which produces a pile upon hardening.

When a pile foundation is decided upon, it is necessary to compute the required pile cross section and length based on the load from the superstructure, allowable stress in the pile material (usually a code value), and the in situ soil properties. ② These requirements allow the foundation contractor to order the necessary number and lengths of piles. Dynamic formulas, pile-load tests, or a combination are used on-site to determine if the piles are adequately designed and placed. It is generally accepted that a load test is the most reliable means of determining the actual pile capacity.

Determination of pile bearing capacity are very difficult. A large number of different equations are used, and seldom will any two give the same computed capacity. Organizations that have been using a particular equation tend to stick with it—particularly if a successful data base has been established. In a design situation one might compute the pile capacity by several equations using the required empirical factors suitably adjusted (or estimated) and observe the computed capacity. From a number of these computations some "feel" for the probable capacity will develop so that a design recommendation/proposal can be made.

Words and Expressions

transmit	[trænz'mit]	*v.* 发射,播送,传递
uplift	[ʌp'lift]	*n.* &*v.* 提高,振奋
overturn	[ˌəuvə'təːn]	*v.* 翻倒,倾覆
basement	['beismənt]	*n.* 地下室,地下层
cohesionless	[kəu'hiːʒənlis]	*adj.* 非黏结性的
scour	['skɑuə]	*v.* 冲刷,擦净,擦亮,冲刷成 *n.* 冲刷,擦
offshore	['ɔf 'ʃɔː]	*adj.* 近海的,向海的,离岸的
embed	[im'bed]	*v.* 把……嵌入,埋入
landslide	['lændslaid]	*n.* 山崩,(山坡、悬崖等的)崩塌,滑坡
parallel	['pærəlel]	*adj.* 平行的,类似的,相对应的
recommendation	[ˌrekəmen'deiʃən]	*n.* 推荐,劝告,建议
pile shaft		桩身
friction pile		摩擦桩
side resistance		侧阻力
point bearing		端阻力
water table		水位

tower leg	塔支柱
amplitude of vibration	振幅
natural frequency	自振频率
power pole	电线杆
pile hammer	桩锤
single pile	单桩
pile group	群桩
tension pile	受拉桩
allowable stress	容许应力
dynamic formula	动力打桩公式
pile-load test	桩荷载试验
empirical factor	经验系数

Notes

① In any case great care should be exercised in determining the soil properties at the site for the depth of possible interest so that one can as accurately as possible determine whether a pile foundation is needed and, if so, that neither an excessive number nor lengths are specified.
应特别仔细地确定场地可能的影响深度内土体的特性，以便尽可能精确地决定是否需要采用桩基，这样就不至于设计出过多的桩数和过长的桩长。

② When a pile foundation is decided upon, it is necessary to compute the required pile cross section and length based on the load from the superstructure, allowable stress in the pile material (usually a code value), and the in situ soil properties.
当决定采用桩基时，有必要根据上部结构荷载、桩身材料的容许应力(通常为规范值)和原位土体参数来计算所需的桩截面尺寸和桩长。

Exercises

Ⅰ. Translate the following words into Chinese.

1. pile group ____________________
2. tension pile ____________________
3. skin resistance ____________________
4. natural frequency ____________________
5. dynamic formula ____________________

Ⅱ. Translate the following words into English.

1. 桥墩____________________
2. 端承桩____________________
3. 容许应力____________________
4. 振幅____________________

Ⅲ. Translate the following sentences into Chinese.

1. This distinction is purely one of convenience since all piles carry load as a combination of side resistance and point bearing except when the pile penetrates an extremely soft soil to a solid base.
2. To resist uplift, or overturning, forces, such as for basement mats below the water table or to support tower legs subjected to overturning from lateral loads such as wind.
3. In those cases where piles are used to control settlement at marginal soil sites, care should be taken to utilize both the existing ground and the piles in parallel so that a minimum number are required.
4. This produces both considerable noise and local vibrations, which may be disallowed by local codes or environmental agencies and, of course, may damage adjacent property.
5. In a design situation one might compute the pile capacity by several equations using the required empirical factors suitably adjusted (or estimated) and observe the computed capacity.

Ⅳ. Translate the following sentences into English.

1. 桩是用来将表面荷载传递到土体中较深位置的木、混凝土或钢的结构构件。
2. 通过桩体排土和打入振动的共同作用压缩松软、无黏性沉积层,这种桩以后可以拔出。
3. 在近海岸的结构物中,可将水面以上的荷载穿过水传递到下卧土层中。
4. 桩基比扩展基础的成本高很多,且可能比筏基昂贵。
5. 现场使用动力打桩公式、桩承载力试验或者两者的结合来确定设计和桩位是否恰当。

Ⅴ. Answer the following questions briefly.

1. What is the purpose of pile? Try to state it in your own words.
2. What is the characteristic of load distribution for pile subjected to the vertical load?
3. How many methods are there to determine pile capacity?

Reading Materials

Improving Site Soils for Foundation Use

The centuries-old problem of land scarcity in the vicinity of existing urban areas often necessitates the use of sites with soils of marginal quality. In many cases these sites can be utilized for the proposed project by using some kind of soil improvement. An extremely large number of methods have been used and/or reported in the literature—many of which have been parented—and at an individual site one may use a mix of several methods to achieve the desired result.

For a given site a first step is to make a literature review of at least some of the methods reported. This together with a reasonable knowledge of geotechnical fundamentals allows the engineer to use either an existing method, a mix of methods, or some methods coupled with modest *ingenuity*(机灵的,独创性的) (unless limited by a governmental agency) to produce an adequate solution for almost any site.

Of principal interest in this article is the identification of means to obtain a significant increase in the bearing capacity of a soil. This can be achieved by altering the soil properties of ϕ, cohesion c, or density ρ. Usually an increase in density (or unit weight γ) is accompanied by an increase in either ϕ or c or both (assuming the soil is cohesive). *Particle*(粒子) packing (compaction) always increases the density, with a resulting decrease in void ratio, and reduces long-term settlements. Particle packing usually increases the stress-strain modulus so that any "immediate" settlements are also reduced.

Mechanical stabilization. In this method the grain size gradation of the site soil is altered. Where the site soil is predominantly gravel (say, from 75 mm down to 1 mm), *binder*(黏合剂) material is added. Binder is defined as material passing either the No. 40 (0.425 mm) or No. 100 (0.150 mm) sieve. The binder is used to fill the voids and usually adds mass cohesion. Where the soil is *predominantly*(卓越的) *cohesive*(黏着的) (No. 40 and smaller sieve size), *granular*(由小粒而成的) soil is imported and blended with the site soil.

In either case the amount of improvement is usually determined by trial, and experience shows that the best improvement results when the binder (or filler) occupies between 75 and 90 percent of voids of the coarse material. It usually requires much more granular materials to stabilize cohesive deposits than binder for cohesionless deposits and as a result other stabilizing methods are usually used for clayey soils.

Compaction. This method is usually the most economical means to achieve particle packing for both cohesionless and cohesive soils and usually uses some kind of rolling equipment. Dynamic compaction is a special type of compaction consisting of dropping heavy weights on the soil.

Preloading. This step is taken primarily to reduce future settlement but may also be used to increase shear strength. It is usually used in combination with drainage.

Drainage. This method is undertaken to remove soil water and to speed up settlements under preloading. It may also increase shear strength since s_u, in particular, depends on water content. For example, consolidation without drainage may take several years to occur whereas with drainage facilities installed the consolidation may occur in 6 to 12 months.

Densification using *vibratory*(振动的)equipment. Densification is particularly useful in sand, silty sand, and gravelly sand deposits with D_r less than about 50 to 60 percent. This method uses some type of vibrating *probe*(探针，探测器), which is inserted into the soil mass and withdrawn. Quality fill is added to the site to bring the soil surface to the required grade since the site soil usually settles around and in the vicinity of the vibrating probe.

Use of in situ reinforcement. This approach is used with stone, sand, cement, or lime columns. This treatment produces what is sometimes called composite ground. Sometimes small amounts of short lengths of plastic fibers or fiberglass can be mixed with the soil for strength improvement. The major precaution is to use a fiber material that has an adequate durability in the hostile soil environment.

Grouting. Initially this was the name for injection of a viscous fluid to reduce the *void ratio*(孔隙率) (and k) or to cement rock cracks [see ASCE (1962)]. Currently this term is loosely used to describe a number of processes to improve certain soil properties by injection of a viscous fluid, sometimes mixed with a volume of soil. Most commonly, the viscous fluid is a mix of water and cement or water and lime, and/or with additives such as fine sand, *bentonite*(斑脱土) clay, or fly ash. Bitumen and certain chemicals are also sometimes used. *Additives*(添加剂) are used either to reduce costs or to enhance certain desired effects. Since the term grout is so loosely used in construction, the context of usage is important to define the process.

Use of *geotextiles*(土工织物). These function primarily as reinforcement but sometimes in other beneficial modes.

Chemical stabilization. This means of stiffening soil is seldom employed because of cost. The use of chemical stabilizers is also termed chemical grouting. The more commonly used chemical agents are *phosphoric acid*(磷酸), *calcium chloride*(氯化钙), and *sodium silicate*(硅酸钠)(or water glass). Some laboratory tests indicate certain metallic powders (aluminum, iron) may produce beneficial effects as well [Hoshiya and Mandal (1984)]. ASCE (1957, 1966) cited usage of an extremely large number of chemical grouting procedures (mostly patented, but most of the patents have probably expired by now).

Strictly, soil-cement and lime-soil treatment (often together with fly ash and/or sand) is a chemical stabilization treatment, but it is usually classified separately.

科技英语写作技巧(8)——“例证、举例”的表示方法

常用词语结构：

as an example/illustration 作为(一个)……的例子

be an example of＋NP 是……的一个例子
be exemplified by＋NP ……的例子是……
by way of example/illustration 作为例子
for example/instance（e. g. ）例如
take…as/for an example 以……为例
as follows 如下
such as 像……这样的，例如……

例句：

➢ The experiment，for example，was a total failure.
例如，那次实验是完全不成功的。

Lesson 23

RETAINING WALL

Retaining walls are used to prevent retained material from assuming its natural slope. Wall structures are commonly used to support earth, coal, ore piles, and water. Most retaining structures are vertical or nearly so; however, if the α angle in the Coulomb earth-pressure coefficient is larger than 90°, there is a reduction in lateral pressure that can be of substantial importance where the wall is high and a wall tilt into the backfill is acceptable. ①

Retaining walls may be classified according to how they produce stability.

① Mechanically reinforced earth—also sometimes called a "gravity" wall.

② Gravity—either reinforced earth, masonry, or concrete.

③ Cantilever—concrete or sheet-pile.

④ Anchored—sheet-pile and certain configurations of reinforced earth.

At present, tile mechanically stabilized earth and gravity walls are probably the most used—particularly for roadwork where deep cuts or hillside road locations require retaining walls to hold the earth in place. ② These walls eliminate the need for using natural slopes and result in savings in both right-of-way costs and fill requirements.

Cantilever walls of reinforced concrete are still fairly common in urban areas because they are less susceptible to vandalism and often do not require select backfill. Typically they compete well in costs where the wall is short (20 to 50 m in length) and not very high (say, under 4 m). They are also widely used for basement walls and the like in buildings.

The mechanically reinforced earth wall of Fig. 23-1 uses the principle of placing reinforcing into the backfill using devices such as metal strips and rods, geotextile strips and sheets and grids, or wire grids. There is little conceptual difference in reinforcing soil or concrete masses—reinforcement carries the tension stresses developed by the applied loads for either material. Bond stresses resist rebar pullout in concrete; soil relies on friction stresses developed based on the angle of friction δ between soil and reinforcement or a combination of friction and passive resistance with soil and wire grids.

The principle of reinforced earth is not new. Straw, bamboo rods, and similar alternative materials have long been used in technologically unsophisticated cultures to reinforce mud bricks and mud walls. ③ Nevertheless, in spite of this long usage,

French architect H. Vidal was able to obtain a patent (ca. mid-1960s) on the general configuration of Fig. 23-1, which he termed "reinforced earth". We see three basic components in this figure.

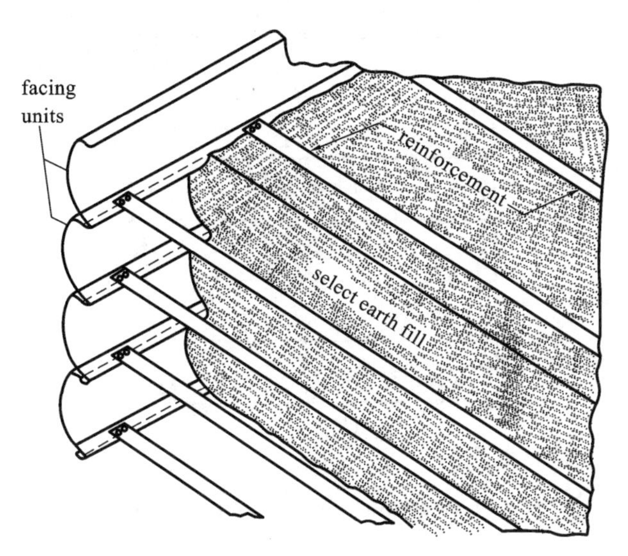

Fig. 23-1 The reinforced earth concept

① The earth fill—usually select granular material with less than 15 percent passing the No. 200 sieve.

② Reinforcement—strips or rods of metal, strips or sheets of geotextiles, wire grids, or chain link fencing or geogrids (grids made from plastic) fastened to the facing unit and extending into the backfill some distance. Vidal used only metal strips.

③ Facing unit—not necessary but usually used to maintain appearance and to avoid soil erosion between the reinforcements.

These three components are combined to form a wall whose side view is shown in Fig. 23-2. When wire mesh or other reinforcement with discontinuities (grid voids) is used, a portion may be bent, similar to the sheet of Fig. 23-3, to form a facing unit. Grid-type reinforcements strengthen the soil through a combination of friction and passive pressure pullout resistance. The bent-up portion used as a facing piece provides some erosion control until the wall is completed.

The exposed reinforcements are usually sprayed with concrete mortar or gunite (material similar to mortar) in lifts to produce a thickness on the order of 150 to 200 mm. This is both to improve the appearance and to control erosion. For metals this covering also helps control rust, and for geotextiles it provides protection from the ultraviolet rays in sunlight and discourages vandalism.

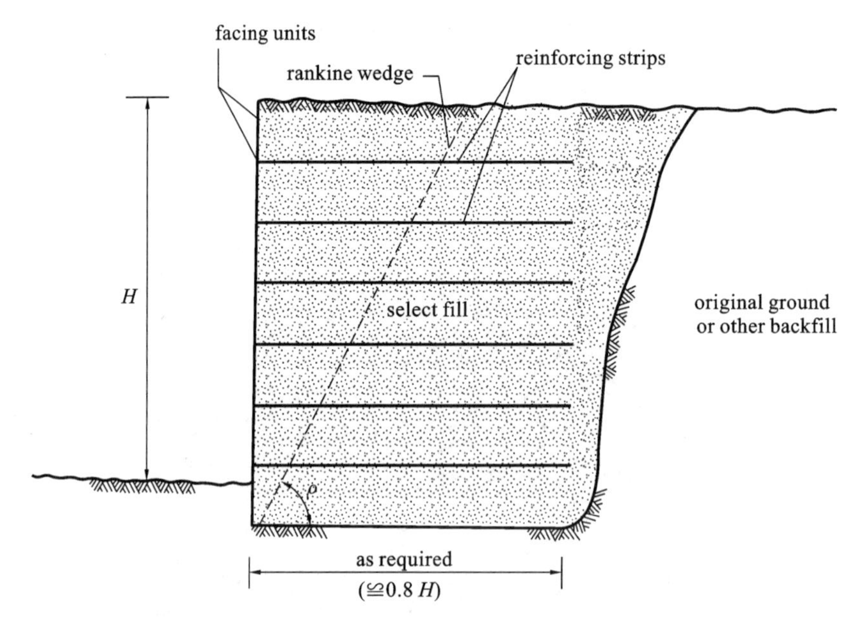

Fig. 23-2 Line detail of a reinforced earth wall in place

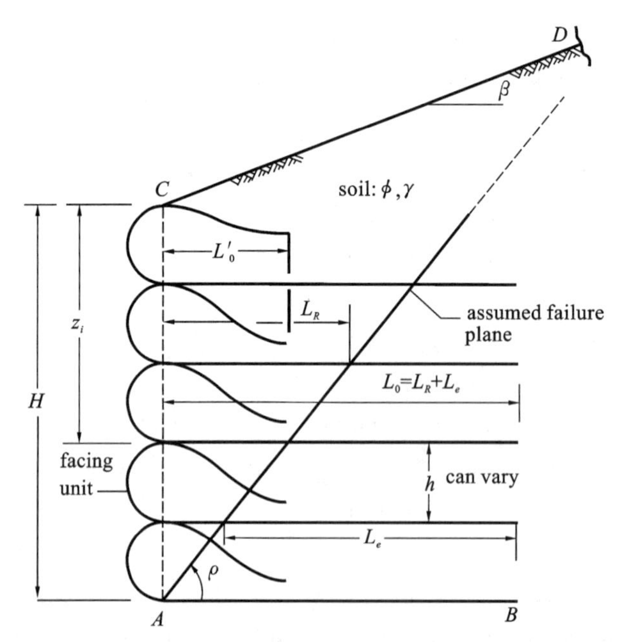

Fig. 23-3 Using geotextile sheets for reinforcement with the facing unit formed by lapping the sheet as shown

The basic principle of reinforced earth is shown in Fig. 23-4 where we see a wall acted on by either the Rankine or Coulomb active earth wedge. Full-scale tests have verified that the earth force developed from the active earth wedge at any depth z is carried by reinforcing strip tension.

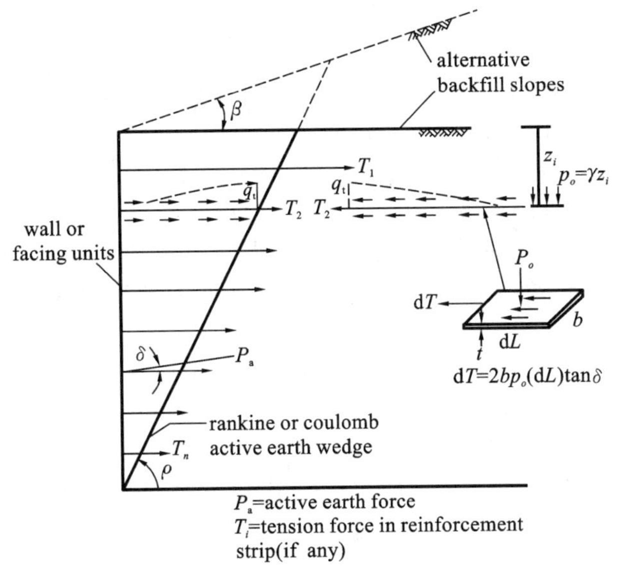

Fig. 23-4 The general concept of reinforced earth is that $\sum T_i = P_a \cos\delta$, so the earth force against the wall(or facing units) = 0

Strip tension is developed in the zone outside the active earth wedge from the friction angle δ between strip and soil and the vertical earth pressure γz on the strip. With no lateral earth pressure left to be carried by the wall facings, they can be quite thin and flexible with the principal functions of erosion control and appearance.

Words and Expressions

retain	[riˈtein]	*v.* 保持，保留
backfill	[ˈbækfil]	*v.* 回填
cantilever	[ˈkæntiliːvə]	*n.* 悬臂，悬臂梁
roadwork	[ˈrəudwəːk]	*n.* 长跑运动训练，道路作业
vandalism	[ˈvændəlizəm]	*n.* 肆意破坏公物等行为
conceptual	[kənˈseptjuəl]	*adj.* 观念的，概念的
geogrid	[gjəuˈgrid]	*n.* 格栅，土工格栅
gunite	[ˈgʌnait]	*n.* 压力喷浆，喷射的水泥砂浆
ultraviolet	[ˌʌltrəˈvaiəlit]	*adj.*（光）紫外线的
coulomb earth-pressure		库伦土压力
lateral pressure		侧压力
mechanically reinforced earth		加筋土
sheet-pile		板桩
gravity wall		重力式挡墙
right-of-way		通行权，筑路权
full-scale test		足尺试验

Notes

① Most retaining structures are vertical or nearly so; however, if the α angle in the Coulomb earth-pressure coefficient is larger than 90°, there is a reduction in lateral pressure that can be of substantial importance where the wall is high and a wall tilt into the backfill is acceptable.
大多数挡土墙是竖直的或者接近竖直的,然而,当库伦土压力系数中的 α 角大于90°时,侧压力将减少,这对于高度高的挡土墙以及向回填区倾斜的挡土墙具有重要意义。

② At present, the mechanically stabilized earth and gravity walls are probably the most used—particularly for roadwork where deep cuts or hillside road locations require retaining walls to hold the earth in place.
目前,加筋土挡墙和重力式挡墙广泛应用,尤其是在深挖和山坡位置的道路工程中需要利用挡墙来保持土体稳定的情况下。

③ The principle of reinforced earth is not new. Straw, bamboo rods, and similar alternative materials have long been used in technologically unsophisticated cultures to reinforce mud bricks and mud walls.
加筋土的原理早已出现,并非什么新技术,早期人们长期采用稻草、竹竿以及类似的材料来加强泥砖和泥墙。

Exercises

Ⅰ. Translate the following words into Chinese.

1. retaining wall ______
2. Coulomb earth-pressure ______
3. active earth force ______
4. full-scale test ______
5. mechanically reinforced earth ______

Ⅱ. Translate the following words into English.

1. 重力式挡土墙______
2. 悬臂式挡土墙______
3. 锚锭式挡土墙______
4. 加筋土______
5. 板桩______

Ⅲ. Translate the following sentences into Chinese.

1. These walls eliminate the need for using natural slopes and result in savings in both right-of-way costs and fill requirements.
2. Cantilever walls of reinforced concrete are still fairly common in urban areas because they are less susceptible to vandalism and often do not require select

backfill.

3. The mechanically reinforced earth wall of Fig. 23-1 uses the principle of placing reinforcing into the backfill using devices such as metal strips and rods, geotextile strips and sheets and grids, or wire grids.
4. For metals this covering also helps control rust, and for geotextiles it provides protection from the ultraviolet rays in sunlight and discourages vandalism.
5. With no lateral earth pressure left to be carried by the wall facings they can be quite thin and flexible with the principal functions of erosion control and appearance.

Ⅳ. Translate the following sentences into English.

1. 挡土墙是用来保持土或其他材料稳定性的构筑物，通常用来支撑土体、煤、矿石以及水。
2. 加筋土和混凝土在概念上几乎没有区别——钢筋承受荷载施加于任一材料所产生的拉应力。
3. 拉筋——金属条杆、土工织物、绳网格栅或塑料包装带土工格栅，固定于墙面上并伸至填土中一定长度。
4. 暴露在外的加筋体通常在安装过程中喷射混凝土砂浆或水泥砂浆以形成150～200 mm厚的保护层。
5. 足尺试验表明在主动土楔体面上的任何深度处的土应力由加筋条拉力来承受。

Ⅴ. Answer the following questions briefly.

1. What is a retaining wall? Try to state it in your own words.
2. How are the retaining wall classified?
3. What is the principle of mechanically reinforced earth wall?

Reading Materials

Cantilever Retaining Walls

Fig. 23-5 identifies the parts and terms used in retaining wall design. *Cantilever walls*(悬臂式挡土墙) have these principal uses at present.

① For low walls of fairly short length, "low" being in terms of an exposed height on the order of 0.1 to 3.0 m and lengths on the order of 100 m or less.

② Where the backfill zone is limited and/or it is necessary to use the existing soil as backfill. This restriction usually produces the condition, where the principal wall pressures are from compaction of the backfill in the limited zone defined primarily by the *heel*(墙踵)dimension.

③ In urban areas where appearance and durability justify the increased cost.

In these cases if the existing ground stands without caving for the depth of vertical excavation in order to place (or pour) the wall footing and later the stem, theoretically there is no lateral earth pressure from the existing backfill. The lateral wall pressure produced by the limited backfill zone of width b can be estimated. There is a larger lateral pressure from compacting the backfill (but of unknown magnitude), which may be accounted for by raising the location of the resultant from $H/3$ to 0.4 to 0.5H. Alternatively, use K_o instead of K_a with the $H/3$ resultant location.

It is common for cantilever walls to use a constant wall thickness on the order of 250 mm to seldom over 300 mm. This reduces the labor cost of form setting, but some overdesign should be used so that the lateral pressure does not produce a tilt that is obvious—often even a few millimeters is noticeable.

Fig. 23-5 gives common dimensions of a cantilever wall that may be used as a guide in a hand solution. Since there is a substantial amount of busywork in designing a retaining wall because of the trial process, it is particularly suited to a computer analysis in which the critical data of γ, φ, H, and a small base width B are input and the computer program iterates to a solution.

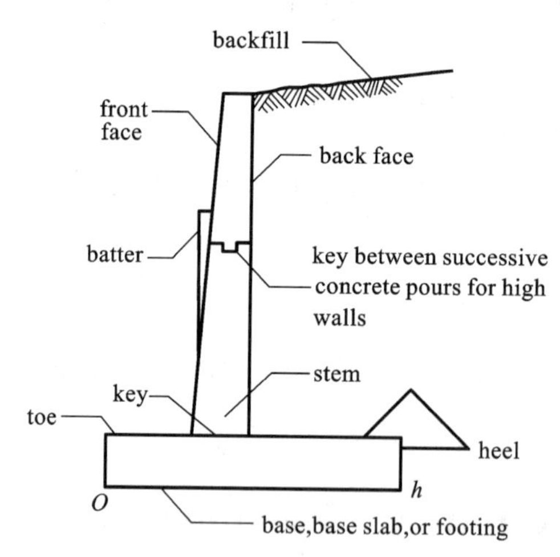

Fig. 23-5 Principal terms used with retaining walls

The dimensions of Fig. 23-6 are based heavily on experience accumulated with stable walls under Rankine conditions. Small walls designed for lateral pressures from compaction, and similar, may produce different dimensions.

It is common, however, for the base width to be on the order of about 0.5H, which depends somewhat on the *toe*(墙脚) distance ($B/3$ is shown, but it is actually not necessary to have any toe). The thickness of the stem and base must be adequate for wide-beam shear at their intersections. The stem top thickness must be adequate for temperature-caused spalls and impacts from equipment/automobiles so that if a piece chips off, the remainder appears safe and provides adequate clear reinforcement cover.

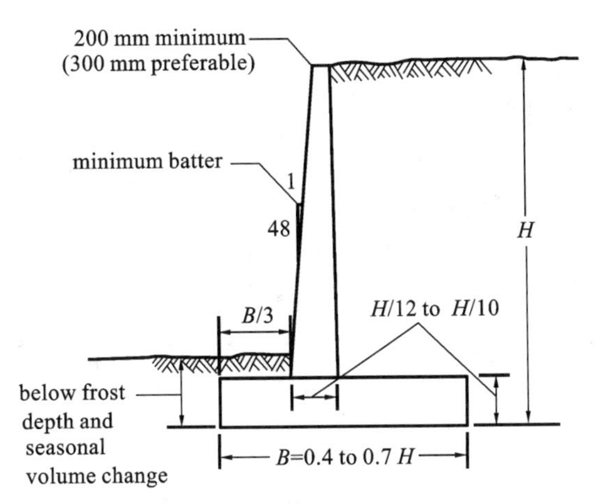

Fig. 23-6 ***Tentative*****（试验性的）design dimensions for a cantilever retaining wall**

Walls are designed for wide-beam shear with critical locations as indicated in ACI 318. The author suggests, however, taking the wide-beam shear at the stem face (front and back) for the base slab as being more conservative and as requiring a negligible amount of extra concrete. For the stem one should take the critical wide-beam location at the top of the base slab. The reason is that the base is usually poured first with the stem reinforcement set. Later the stem forms are set and poured, producing a discontinuity at this location.

Formerly, a wood strip was placed into the base slab and then removed before the stem was placed. This slot or key provided additional shear resistance for the stem, but this is seldom done at present. Without the key at this discontinuity, the only shear resistance is the bonding that develops between the two pours＋any friction from the stem weight＋reliance on the stem reinforcement for shear.

科技英语写作技巧(9)——“例如、排除”的表示方法

常用词语结构：

except＋NP 除……之外

except for＋NP 除……之外

except that… 除……之外，只是……

with the exception of＋NP 除……之外

with the exception that… 除……之外，例外的一点是……

excluding＋NP 不包括……在内

NP＋excluded 不包括……

exclusive of＋NP 不计算在内，……除外

but＋NP 除了……

other than… 除了……

without+NP 而不……

otherwise 其他方面

表示"例如、排除"的例句:

➢ The total cost of the construction, exclusive of material costs, would be…

工程的总造价(材料费不计在内)为……

➢ Everything went as expected with the exception of one small anomaly.

除了小小异常之处外,其他一切都按预料的进行。

科技英语写作技巧(10)——"方式、方法、手段"的表示方法

常用词语结构:

as+-ed/ 分句 像……(那样)

by+NP/-ing 用……,靠……;通过……,用……方法

by means of+NP 借助于……

by/through the application of+NP 通过(靠)施加……

by/through the use of+NP 通过使用……,利用……

by this method/in this way 这样,用这个方法

in a … fashion/manner/way 以……的方式(方法)

in a way that/in which… 以……的方式

in such a way/manner/fashion that… 以这样的方式,即(以至)……

like+NP 像……(那样)

so(+-ed)+that… 这样……以致……

so+-ed+as to+*v*. 这样……以致……

through+NP/-ing 通过……,经由……,以……

whereby+分句(by means of which…)靠那个,借助于那个……

with+NP 用……,以……

with the help/aid of+NP 借助于……,在……帮助下

Lesson 24

CONSTRUCTION ENGINEERING

Construction engineering is a specialized branch of civil engineering concerned with the planning, execution, and control of construction operations for such projects as highways, buildings, dams, airports, and utility lines.

Planning consists of scheduling the work to be done and selecting the most suitable construction methods and equipment for the project. Execution requires the timely mobilization of all drawings, layouts, and materials on the job to prevent delays to the work. Control consists of analyzing progress and cost to ensure that the project will be done on schedule and with the estimated cost.

Planning

The planning phase starts with a detailed study of construction plans and specifications. From this study, a list of all items of work is prepared, and related items are then grouped together for listing on a master schedule. A sequence of construction and the time to be allotted for each item is then indicated. The method of operation and the equipment to be used for the individual work items are selected to satisfy the schedule and the character of the project at the lowest possible cost.

The amount of time allotted for a certain operation and the selection of methods of operation and equipment is readily available to the contractor. After the master or general construction schedule has been drawn up, subsidiary detailed schedules or forecasts are prepared from the master schedule. These include individual schedules for procurement of materials, equipment, and labor as well as forecasts of cost and income.

Execution

The speedy execution of the project requires the ready supply of all materials, equipment, and labor when needed. The construction engineer is generally responsible for initiating the purchase of most construction materials and expediting their delivery to the project. Some materials, such as structural steel and mechanical equipment, require partial or complete fabrication by a supplier. For these fabricated materials the engineer must prepare or check all fabrication drawings for accuracy and case of assembly and often inspect the supplier's fabrication.

Other construction engineering duties are the layout of the work by surveying methods, the preparation of detail drawings to clarify the design engineer's drawings for the construction crews, and the inspection of the work to ensure that it complies with plans and specifications.

For most large projects it is necessary to design and prepare construction drawings for temporary construction facilities, such as drainage structures, access roads, office and storage buildings, formworks, and cofferdams. Other problems are the selection of electrical and mechanical equipment and the design of structural features for concrete material processing and mixing plants and for compressed air, water, and electrical distribution systems. ①

Control

Progress control is obtained by comparing actual performance on the work against the desired performance set up on the master or detailed schedules. ② Since delay on one feature of the project could easily affect the entire job, it is often necessary to add equipment or crews to speed up the work.

Cost control is obtained by comparing actual unit cost for individual work items against estimated or budgeted unit cost, which are set up at the beginning of the work. Unit cost is obtained by dividing the total cost of an operation by the number of units in that operation.

Typical units are cubic yards for excavation or concrete work and tons for structural steel. The actual unit cost for any item at any time is obtained by dividing the accumulated cost charged to that item by the accumulated units of work performed.

Individual work items cost is obtained by periodically distributing job cost, such as payroll and invoices to the various work item accounts. Payroll and equipment rental charges are distributed with the aid of time cards prepared by crew foreman. The cards indicate the time spent by the job crews and equipment on the different elements of the work. The allocation of material cost is based on the quantity of each type of material used for each specific item.

When the comparison of actual and estimated unit cost indicates an overrun, an analysis is made to pinpoint the cause. If the overrun is in equipment cost, it may be that the equipment has insufficient capacity or that it is not working properly. If the overrun is in labor cost, it may be that the crews have too many men, lack proper supervision, or the project is being delayed for lack of materials or layout. In such cases time studies are invaluable in analyzing productivity.

Construction operations are generally classified according to specialized fields. These include preparation of the project site, earthmoving, foundation treatment, steel erection, concrete placement, asphalt paving. Procedures for each of these fields are generally the same, even when applied to different projects, such as buildings, dams, or airports. However, the relative importance of each field is not the same in all cases.

Preparation of site

This consists of the removal and clearing of all surface structures and growth from the site of the proposed structure. A bulldozer is used for small structures and trees. Larger structures must be dismantled.

Earthmoving

This includes excavation and the placement of earth fill. Excavation follows preparation of the site, and is performed when the existing grade must be brought down to a new elevation. Excavation generally starts with the separate stripping of the organic topsoil, which is later reused for landscaping around the new structure. This also prevents contamination of the nonorganic material which may be required for fill. Excavation may be done by any of several excavators, such as shovels, draglines, clamshells, cranes, and scrapers.

Efficient excavation on land requires a dry excavation area, because many soils are unstable when wet and cannot support excavating and hauling equipment. Dewatering becomes a major operation when the excavation lies below the natural water table and intercepts the groundwater flow. When this occurs, dewatering and stabilizing of the soil may be accomplished by well-point and electro osmosis.

Some materials, such as rock, cemented gravels, and hard clays, require blasting to loosen or fragment. Blast holes are drilled in the materials; explosives are then placed in the blast holes and detonated. The quantity of explosives and the blast-hole spacing are dependent upon the type and structure of the rock and diameter and depth of the blast holes.

After placement of the earth fill, it is almost always compacted to prevent subsequent settlement. Compaction is generally done with sheep's-foot, grid, pneumatic-tired, and vibratory-type rollers, which are towed by tractors over the fill as it is being placed. Hand-held, gasoline-driven rammers are used for compaction close to structures where there is no room for rollers to operate.

Foundation treatment

Where subsurface investigation reveals structural defects in the foundation area to be used for a structure, the foundation must be strengthened. Water passages, cavities, fissures, faults, and other defects are filled and strengthened by grouting. Grouting consists of injection of fluid mixtures under pressure. The fluids subsequently solidify in the voids of the strata. Most grouting is done with cement and water mixtures but other mixture ingredients are asphalt, cement and clay, and precipitating chemicals.

Steel erection

The construction of a steel structure consists of the assembly at the site of mill-rolled or shop-fabricated steel sections. The steel sections may consist of beams, columns, or small trusses which are joined together by riveting, bolting, or welding. It is more economical to assemble sections of the structure at a fabrication shop rather than in the field, but the size of preassembled units is limited by the capacity of transportation and erection equipment. The crane is the most common type of erection equipment, but when a structure is too high or extensive in area to be erected by a crane, it is necessary to place one or more derricks on the structure to handle the steel. In high structures the derrick must be constantly dismantled and reelected to successively higher levels to raise the structures. For river bridges, the steel may be handled by cranes on barges, or, if the bridges is too high, by traveling derricks which ride on the bridge being erected. Cables for long suspension bridges are assembled in place by special equipment that pulls the wire from a reel, set up at one anchorage, across to the opposite anchorage, repeating the operation until the bundle of wires is of the required size.

Concrete construction

Concrete construction consists of several operations: forming, concrete production, placement, and curing. Forming is required to contain and support the fluid concrete within its desired final outline until it solidifies and can support itself. The form is made of timber or steel sections or a combination of both and is held together during the concrete placing by external bracing or internal ties. The forms and ties are designed to withstand the temporary fluid pressure of the concrete.

The usual practice for vertical walls is to leave the forms in position for at least a day after the concrete is placed. They are removed when the concrete has solidified or set. Slip-forming is a method where the form is constantly in motion, just ahead of the level of fresh concrete. The form is lifted upward by means of jacks which are mounted on vertical rods embedded in the concrete and are spaced along the perimeter of the structure. Slip forms are used for high structures such as silos, tanks, or chimneys.

Concrete may be obtained from commercial batch plants which deliver it in mix trucks if the job is close to such a plant, or it may be produced at the job site.③ Concrete production at the job site requires the erection of a mixing plant, and of cement and aggregate receiving and handling plants. Aggregates are sometimes produced at or near the job site. This requires opening a quarry and erecting processing equipment such as crushers and screens.

Concrete is placed by chuting directly from the mix truck, where possible, or from buckets handled by means of cranes or cableways, or it can be pumped into place by special concrete pumps.

Curing of exposed surfaces is required to prevent evaporation of mix water or to replace moisture that does evaporate. The proper balance of water and cement is required to develop full design strength.

Concrete paving for airports and highways is a fully mechanized operation. Batches of concrete are placed between the road forms from a mix truck or a movable paver, which is a combination mixer and placer. A series of specialized pieces of equipment, which ride on the forms, follow to spread and vibrate the concrete, smooth its surface, cut contraction joints, and apply a curing compound.

Asphalt paving

This is an amalgam of crushed aggregate and a bituminous binder. It may be placed on the roadbed in separate operations or mixed in a mix plant and spread at one time on the roadbed. Then the pavement is compacted by rollers.

Words and Expressions

execution	[ˌeksiˈkjuːʃən]	*n.* 执行,完成,实施
master	[ˈmɑːstə]	*adj.* 主要的,总的,熟练的
subsidiary	[səbˈsidjəri]	*adj.* 辅助的,补充的,次要的
clarify	[ˈklærifai]	*v.* 澄清,阐明,净化,解释
budget	[ˈbʌdʒit]	*n.* 预算
		v. 做预算,编入预算
payroll	[ˈpeirəul]	*n.* 发放工资额,工资单
invoice	[ˈinvɔis]	*n.* 发票,发货单
overrun	[ˌəuvəˈrʌn]	*v.* 超过,超出,超限
productivity	[ˌprɔdʌkˈtiviti]	*n.* 生产力,劳动生产率
bulldozer	[ˈbulˌdəuzə]	*n.* 推土机,开土机
dismantle	[disˈmæntl]	*v.* 拆除,拆卸
contamination	[kənˌtæmiˈneiʃən]	*n.* 玷污,污染,污染物
shovel	[ˈʃɔvəl]	*n.* 铲,挖掘机,单斗挖土机
dragline	[ˈdræglain]	*n.* 拉索,拉铲挖土机
clamshell	[ˈklæmʃəl]	*n.* 抓斗
scraper	[ˈskreipə]	*n.* 铲运机,刮土机,平土机
haul	[hɔːl]	*v.* 拖,拉,用力拖拉,拖运
fragment	[ˈfrægmənt]	*n.* 断片,碎块,凝固
solidify	[səˈlidifai]	*v.* 固化,固结,凝固
ingredient	[inˈgriːdiənt]	*n.* 组分,成分,配料
derrick	[ˈderik]	*n.* 起重机,(钻井)井口上的铁架塔
silo	[ˈsailəu]	*n.* 筒仓,竖井,(导弹)发射井
amalgam	[əˈmælgəm]	*n.* 混合物,软的混合物
earthmoving		土方工程;大量掘土(或运土)
foundation treatment		地基处理

steel erection	钢结构架设
concrete placement	混凝土浇筑
well-point	降低地下水位的井点,深坑点
slip form	滑动模板,滑模(施工法)

Notes

① Other problems are the selection of electrical and mechanical equipment and the design of structural features for concrete material processing and mixing plants and for compressed air, water, and electrical distribution systems.
and 作为并列连词,一般用在几个并列成分的最后两个之间。本句中,and 出现在所有并列成分之间,这在修辞学上称为连词叠用。全句意思是:其他的问题是选择电气与机械设备和混凝土原料加工的设计与搅拌厂以及压缩空气、配水、配电系统。

② Progress control is obtained by comparing actual performance on the work against the desired performance set up on the master or detailed schedules.
compare…against 为动词与介词搭配。全句意思是:进度控制是通过比较工程实际进度与主要(或详细)进度表中确定的预期进度来进行的。

③ Concrete may be obtained from commercial batch plants which deliver it in mix trucks if the job is close to such a plant, or it may be produced at the job site.
or 表示选择的并列结构,意思为"否则"。全句意思是:如果工地离商品混凝土搅拌厂很近,混凝土就可以从用搅拌车运送混凝土的商品混凝土搅拌厂获得,否则可以在工地制作。

Exercises

Ⅰ. Translate the following words into Chinese.

1. construction operations ____________
2. on schedule ____________
3. construction plans and specifications ____________
4. partial or complete fabrication by a supplier ____________
5. temporary construction facilities ____________
6. proper balance of water and cement ____________

Ⅱ. Translate the following words into English.

1. 商品混凝土搅拌厂____________
2. 材料的采购____________
3. 建筑物周边____________

Ⅲ. Translate the following sentences into Chinese.

1. It is more economical to assemble sections of the structure at a fabrication shop rather than in the field, but the size of preassembled units is limited by

the capacity of transportation and erection equipment.

2. Concrete is placed by chuting directly from the mix truck, where possible, or from buckets handled by means of cranes or cableways, or it can be pumped into place by special concrete pumps.
3. The construction engineer is generally responsible for initiating the purchase of most construction materials and expediting their delivery to the project. Some materials, such as structural steel and mechanical equipment, require partial or complete fabrication by a supplier.
4. When the comparison of actual and estimated unit cost indicates an overrun, an analysis is made to pinpoint the cause.

Ⅳ. Translate the following sentences into English.

1. 计划包含工作的进度安排和选择最适宜于工程的施工方法和设备。
2. 成本控制通过比较单项工程实际单位成本和工程开始时制定的单位成本的估算或预算来实现。
3. 地质勘察显示供结构使用的区域地基有缺陷的地方，基础必须加强。
4. 当挖土位于自然水位以下并截断地面水流时，降水成为一个主要的工作。

Ⅴ. Answer the following questions briefly.

1. How can cost control be obtained?
2. What does concrete construction consist of?
3. On most large projects, which temporary construction facilities are necessary to design and prepare construction drawings for?

Reading Materials

The Trends of Construction IT in the USA

The editors of *Engineering News & Report* (ENR) surveyed readers and interviewed industry leaders and gave a report in its June 21, 2004 issue on 10 Electronic technologies that changed construction since early 1970s. The 10 electronic technologies are the internet, computer aided design, laser, analysis software, personal computers, the fax, critical path method, calculators, mobile communications, global positioning system.

One can trace the roots of these 10 technologies back to the background technologies catalyzing the rapid development of IT. Each technology needs hardware, software, and/or system integration.

Latest development

In author's view, the most significant IT's impact on construction in the last decade is the Internet, CAD, and GPS.

(1) Internet.

Internet was originally used as an e-communication tool for mailing internally in most big US construction companies. Nowadays, Internet has proliferated into many applications in construction.

As ENR pointed out, "Along with allowing people to browse the web, send e-mail, upload and download files of all sorts, the Internet allows people to collaborate, even if their geographical location, hours, language or requirements are different." Web enables construction people to do many things. Web/World Wide Web (WWW) enables construction personnel to navigate the network through graphical user interface and to hyperlink to the world inside and/or outside of the company where the user works for. Web allows AEC to get e-permits, to make business transactions through e-commerce, to customize parts on e-Marchineshop, to conduct distance learning on line, and to perform web-enabled project management functions and so forth. Moreover, many construction software applications are put on platforms accessible to anyone in the system at any time and any place.

(2)3D CAD/4D/VR.

Many AEC firms have placed particular value on 3D CAD. The 3D CAD enhances accuracy, completeness, constructability, repeatability, speed, and visualization of the constructed project. Some companies even took one step further by creating interactive software package. The interactive 3D CAD user enables the user to participate by walking through the constructed facility with a dynamic 3D perspective. This gives a feel of how the constructed facility will look, function and interact before any construction occurs.

Visual 3D CAD models combined with scheduling and simulation software display the spatial and temporal aspects of a constructed project simultaneously. This display adds 4th dimension of time on the 3D CAD and is called 4D models. The project manages can base on the 4D model to find the best schedule alternatives for the constructed project.

The CAD modeling has made tremendous strides toward achieving greater compatibility of construction processes in the field. Designers can now use CAD packages to integrate building systems such as electrical, mechanical, plumbing, HVAC, security, and communications to minimize interferences among the various systems and to reduce change orders during construction.

Another innovation provides even more design flexibility is virtual reality (VR). VR is a computer-based, multi-sensory cyberspace of 3D visual images, stereoscopic sounds, and tactile sensations delivered through user interfaces such as goggles and gloves. This is a new type of user interface based on placing the user in a 3D environment that can be directly manipulated. Virtual realities are simulated worlds created by computers loaded with interactive software. Rather than using keyboards to communicate with the machine, people can use voice commands, finger pointing, head nodding, and other natural gestures. The ultimate goal of VR is to simulate a landscape in which one can walk around and manipulate objects at will—a landscape so convincing that the word "virtual" disappears from the term "virtual reality".

Potential construction benefits include using VR to simulate a walk-through of a constructed facility in a full scale 3D dimension and to allow user's participation in the facility before construction has even begun. This would enable a designer, contractor, or owner to actually vision the to-be-built facility. It greatly increases the communication among parties involved and minimizes interferences.

(3)GPS/GIS.

To have a super precise form of worldwide positioning, the U. S. Department of Defense launched constellations of 24 satellites and corresponding ground stations to construct a non-propriety global positioning system (GPS) in 1970s. The GPS can allow the user to pinpoint a location accurate to within a centimeter. Fortunately, the minimization of microprocessor has driven down the cost of GPS receivers to the point at which virtually every one can use it today.

GPS is not only used for positioning the location. It also can provide navigators for cars, ships, and planes to get one location to another, assist the monitoring of the movement of them, and facilitate the surveying and mapping processes, as-built drawing, laying out construction site, controlling heavy earth equipment movement, managements, and so forth. Conjunction with the information provided by GPS such as soil conditions, code requirements, skill level of labors, climate, average daily vehicles, and numerous layer of other information about the location, GPS/GIS combination is becoming a power tools for construction.

On the horizon

(1)Radio frequency identification (RFID)/Smart chips.

Barcodes have been used in construction material management for many years; however, barcodes have limitation in the light-of-sight, difficulty to read multiple items simultaneously, malfunction in harsh environments and inability to write. These limitations lead the use of RFID.

Long-wave radio frequency has been used for many years in telecommunications. More recently, microwave transmission, using the high end of the radio frequency range, has entered use by common carriers, especially on long-haul communication links. This requires less-frequent reamplifications and no wires, permitting very high bandwidth communications.

RFID is consisted of three points: a reader, a transponder and a computer system. The reader broadcasts signal through its antenna. The transponder receives the signal and sends back to an identifying response to reader. And then, the reader sends data to computer system for logging and processing other functions. The transponder is commonly called tag or smart chip.

The ability of RFID tag/smart chip to be read, written and programmed is expected to have significant impact on construction inventory control. RFID can improve the material receiving logs from the point of material being fabricated to point of being installed-in-place. Meanwhile, it enables just-in-time delivery of materials to the construction spot, and shortens the construction duration. Since the RFID tag can carry the historical data of the installed materials and equipments, fabricators/manufactures/ equipment supplies can directly communicate to construction workers with special instructions on installation, inspection, and maintenance. Thus, QA/QC will be better preserved.

RFID-made smart card is contactless. It improves not only the efficiency of controlling the materials, equipments, tools, and personnel, but also, the security and safety of the work places.

Nevertheless, there are challenges ahead for using RFID in reducing the cost, preventing damage on the tags from severe environmental impact, and maintaining the security of the data transmitted.

(2) Handhold/Tablet PCs.

The need of portable PC in construction is evident on construction site. Constructed facility is continuously building up. Construction worker and equipments are moving around constantly. Thus, handheld/tablet PCs are designed to replace the clipboards for notes-taking and diagrams drafting. This characteristic really fits to the construction application. Beyond the checking boxes and filling out forms, handheld/tablet PCs have been developed to the level that handwritten notes can be easily recognized, stored and sorted and retrieved. In addition, the hand-drawn diagrams, charts, and figures can be quickly linked to "Office" documents for further clarification and extension.

The handheld/tablet PCs are expected to be widely used in construction field soon.

(3) Nano-construction.

Looking forward, the recent development of microtechnology has enabled researchers to make, manipulate, and probe objects at the level of a single atom. Researchers also discovered through this enhancement that these tiny atoms and clusters of atoms have completely different properties from those of the same material in bulk form in terms of electrical, magnetic, optical, and mechanical properties.

The discovery of these unique properties has consequently generated a wide variety of innovations in new materials, devices, and sensors. The innovations are expected to impact all sectors of technology from advanced electronics to advanced medicine and are given a common term as "*nanotechnology*(纳米技术)". This is in reference to manipulating and making the matter on atom on a scale of one billionth of a meter—a nanometer. When the feature size of the microprocessor is miniaturized into nano-scale, the nano-processor's speed and memory density will be highly increased while the size, energy needed and cost are continuously reducing. Building on the dramatically improvement of processors through nanotechnology, the construction IT will be profoundly impacted. Likely, the construction industry will walk into nano-construction era after a decade of pursuing e-construction in the next decade.

科技英语写作技巧(11)——"引起、导致、产生"的表示方法

be due to… 是由……所引起的
initiate 触发,引起
bring about/on… 引起……
lead to… 导致……
excite 引起,产生,发展……
produce 产生,引起
cause 引起,产生,发展……
provoke 激起,引起,诱发
develop 引起,激起
raise 引起,唤起
generate 产生,引起
result in… 导致,产生
give rise to… 引起……
set off/up… 引起,激起
induce 引起,导致,(感应)产生

例句:

➢ Operation of the machine sets up/gives rise to vibrations.
机器的运转引起振动。

➢ This may generate/raise a stress.
这样会引起(产生)应力。

Lesson 25

THE PROCEDURE OF THE CIVIL ENGINEERING PROJECT

All civil engineering projects have four distinct phases: ①planning, ②design, ③construction, ④operation.

The relative importance of these phases varies depending on the nature and size of a particular project but all are present to some extent, even in the simplest of projects.[①] A major project such as a new airport or sewage-treatment facility involves many decisions in all phases. It is instructive to examine each of the four phases in turn and to determine the sorts of decisions that have to be made.

The planning phase

The planning phase of a project precedes all the other three phases but can over-lap the design phase. The planning phase starts with the idea for the project and examines that idea from many angles. Perhaps the most important decisions that must be made are those concerned with whether or not to pursue the idea for the project any further. In order to make this decision, many questions must be answered. Among these are the following.

- Is the project needed?
- What will the project cost?
- What will the benefits of the project be?
- Where will it be located?
- How big will it be?
- What impact will it have on the environment?
- Who will pay for the project?
- How will the work be financed?
- What alternatives are available?
- What are their quantitative advantages and disadvantages?

The list is by no means exhaustive. The important thing about these questions is that they are very general yet they each require very specific answers.

Ideally, the objective in planning is to provide the decision-makers with alternatives. Several possible schemes of different sizes and using different methods and concepts should be presented. Each scheme should be technically viable and should have broad total cost estimates associated with it. For each scheme a balance sheet of costs, benefits, advantages and disadvantages should be prepared by the planning group as a whole. Given this information the decision-making body may then make its selection.

For the civil engineering members of a planning group these alternative schemes take the form of feasibility studies. A civil engineering feasibility study itself is an exercise in decision-making. For a project such as the new airport, several different schemes must be evaluated in terms of civil engineering viability and cost. For each scheme many decisions are needed.

How many runways should there be and what length and what orientations should they have?

What terminal facilities, maintenance facilities and car parks should there be and how big should they all be and how should they be laid out?

Some of these decisions encroach on the design phase but at the planning level detailed design is not usually required. Decisions only need to be made such that over-all cost estimates can be drawn up for each scheme. Thus feasibility studies are as much concerned with cost as with technical feasibility. Indeed, almost all schemes can be made technically feasible but sometimes this can only be done at great cost. Feasibility studies are therefore very much exercises in cost modeling. All other things being equal, the least cost scheme usually attracts especial attention.

If several schemes are presented at the planning stage it is important to maintain comparability among them. If one scheme has been worked out in great detail and special efforts have been made to improve and refine it, whereas another scheme has only been very roughly evaluated, the job of choosing one or the other is made more difficult. By how much would the second scheme improve as a result of more care and attention? Ideally, therefore, there should be some common basis for comparison among the alternatives. A basis does exist if a systematic approach to decision-making is used. All participants in the planning process should be motivated by the same desire: to make the best possible decisions. By whatever means this is achieved, formally or informally, mathematically or through experience, this desire is an optimum-seeking motivation.

The design phase

The design phase may be considered to start when the major planning phase decisions have been made. Clearly a decision to proceed with the project must have been made. The other major decisions of the planning phase should have selected a particular scheme from the alternatives and this scheme becomes a framework within which the design phase is conducted. ② Taking the airport for example, the planning decisions should have determined an over-all scheme. A site should be known. Numbers, lengths and orientations of runways should be known. Capacities of terminal facilities for passengers and freight, car-parking capacities, aircraft handling and maintenance capacities should also be known. In other words, a design brief or specification should be available. The design phase consists of providing a complete design that fulfils that specification. Clearly the design phase is a major civil engineering responsibility.

It is helpful to divide the design phase into two parts: macro-design and micro-design. The design brief is usually sufficiently general to allow considerable scope for creative civil engineering design. For example, in the airport scheme the brief may include the provision of underground car-parking for five thousand cars. The macro-design stage must determine how this requirement is to be satisfied. This requires many decisions to be made.

- How big should each one be?
- Where should they be located?
- How many underground stories should there be in each car park?
- What should be the layout of access roads?
- What should be their proximity to terminal facilities?
- What materials should be used?

The decisions that must be made in macro-design are very similar to those for feasibility studies in the planning phase. They narrow down a general requirement to a much more detailed design and produce a very specific brief for the micro-design stage. Micro-design is then concerned with the detailed design of the elements of the project: member sizes, arrangements, joints, etc.

In the macro-design stage major considerations in decision-making are those of cost and technical suitability. Cost models are important for deciding between alternatives but are not the only means available. There are still inputs to the decision-making process from outside the design office. Design decisions cannot be made independently of the means of construction and this implies an overlap with the construction phase. Also there must be consideration given to the users of the designed project. Does it relate well or badly to the over-all project plan as defined by the planning phase? The objectives of macro-design are to make the best possible decisions in the light of all influencing factors—again a process of optimization.

In the micro-design stage the influence of the planning phase is much reduced but the detailed design of all the project elements is influenced by the needs of the construction phase, for example, whether the design will be easy and cheap to build, and by the parallel design stages of non-civil elements. Typically, civil engineering designers must compromise with others over electrical, mechanical, heating and ventilation design requirements. The governing motivation is always to make the best possible decisions. The end product of the design phase should be complete plans and drawings for the entire project. This end product forms the starting brief for the construction phase.

The construction phase

The construction phase turns the project design into reality and the civil engineering contractor is responsible for the construction work. His over-all objective is to complete the construction work within an estimated time, according to the design and generally as efficiently as possible so as to maximize his profits on the contract. ③ In order to do this the contractor must plan his operations very carefully and has many decisions to make.

- What is the best order for the various construction activities?
- How long will each activity take?
- What plant is needed for each activity?
- How many men are needed for each activity?
- How are the men and machinery allocated among activities?
- How much will each activity cost?
- Will the materials be available when required?
- Is there sufficient money available to pay for everything?

Very few of the contractor's decisions are technological ones. Frequently contractors negotiate minor design changes in order to make the construction easier. Also there are usually some unforeseen difficulties, perhaps associated with groundwater or unexpected subsoil properties, etc., which require technological expertise. There is, therefore, some overlap with the design phase. Most of the contractor's decisions, however, concern logistics rather than technology. Furthermore, the contractor has to make each decision very many times over as the construction work alters. Yesterday's allocations of men to tasks must be reviewed and modified to take account of the changed position of the construction work today. The contractor, therefore, is faced with regular intensive decision-making. Most of these decisions are concerned with ensuring that men, machinery, materials and money are available as required and with allocating these resources among the many construction activities which require them.

The operation phase

The last of the four phases of a civil engineering project, the operation phase, is sometimes overlooked yet it is often the most important of the four. All projects have an operation phase; in some it is obvious and important, in others less so. For a new airport or sewage-treatment works the operation phase is the culmination of the project. The first three phases can only be judged to have been successful if the completed project can be operated efficiently. Although the civil engineers concerned with design and construction are not usually actively involved in the operation phase, other civil engineers may be.

Design and construction is usually associated with the private sector of the civil engineering industry. Civil engineering, however, also has a large public sector. Very many civil engineers work for local authorities, water authorities and nationalized industries. The operation phase is of particular interest to this group of civil engineers who are responsible for the running of essential services such as water supply, sewage and solid waste disposal or for the planning and operation of public transport, maintenance of road and rail systems, etc. It is perhaps artificial to class operation as the fourth phase of a project since it usually involves different groups of people but this classification has the merit of providing a reminder that, when all design and construction work is finished, there is still much work to be done by civil engineers.

Words and Expressions

quantitative	[ˈkwɔntitətiv]	*adj.* 数量的,量的
exhaustive	[igˈzɔːstiv]	*adj.* 无遗漏的,彻底的,详尽的
viable	[ˈvaiəbl]	*adj.* 可行的
evaluate	[iˈvæljueit]	*v.* 评价,估计
viability	[ˌvaiəˈbiləti]	*n.* 生存能力,生活能力
comparability	[ˌkɔmpərəˈbiliti]	*n.* 可比较性
freight	[freit]	*n.* 货物,船货,运费,货运
overlap	[ˌəuvəˈlæp]	*v.* 重叠,交叠
negotiate	[niˈgəuʃieit]	*v.* (与某人)商议,谈判
unforeseen	[ˌʌnfɔːˈsiːn]	*adj.* 未预料的
expertise	[ˌekspəˈtiːz]	*n.* 专家的意见,专门技术
logistics	[ləuˈdʒistiks]	*n.* 后勤
overlook	[ˌəuvəˈluk]	*v.* 忽略,没注意到
culmination	[ˌkʌlmiˈneiʃən]	*n.* 顶点
merit	[ˈmerit]	*n.* 优点,价值
by no means		决不
balance sheet		资产负债表
feasibility study		可行性研究
encroach on		侵犯,蚕食
macro-design		宏观设计
narrow down		减少,限制,缩小

Notes

① The relative importance of these phases varies depending on the nature and size of a particular project but all are present to some extent, even in the simplest of projects.

but 用作等立连词，使其前后的词、短语、从句相互对照，作“但是，然而，可是”解。全句意思是：这些阶段的相对重要性取决于项目的特性和规模，但在所有项目中每个阶段都会以某种程度出现，甚至在最简单的项目中也不例外。

② The other major decisions of the planning phase should have selected a particular scheme from the alternatives and this scheme becomes a framework within which the design phase is conducted.
which 用做关系代词，引导限制性定语从句。全句意思是：规划阶段的另一个重要决定是在供选择的方案中选择出一个特定方案，一旦选出方案，这个方案就成为设计阶段进行工作的框架。

③ His over-all objective is to complete the construction work within an estimated time, according to the design and generally as efficiently as possible so as to maximize his profits on the contract.
to… 不定式放在 be 动词后面，形成表语。全句意思是：承包商的总目标是在预定的时间内，按照设计完成施工工作，一般都尽可能高效地完成施工，以获得合同规定的最大利润。

Exercises

Ⅰ. Translate the following words into Chinese.

1. various construction activities ____________________
2. over-lap the design phase ____________________
3. make the actual planning decisions ____________________
4. civil engineers ____________________

Ⅱ. Translate the following words into English.

1. 土木工程项目 ____________________
2. 污水处理厂 ____________________
3. 技术可行 ____________________
4. 施工阶段 ____________________
5. 不可预见的困难 ____________________
6. 设计阶段 ____________________

Ⅲ. Translate the following sentences into Chinese.

1. Design and construction is usually associated with the private sector of the civil engineering industry.
2. Thus feasibility studies are as much concerned with cost as with technical feasibility.
3. Frequently contractors negotiate minor design changes in order to make the construction easier.

Ⅳ. Translate the following sentences into English.

1. 就大型工程而言，也许不是由土木工程师进行总体规划的决策。

2. 施工阶段将工程项目的设计转化为现实,土木工程承包商负责建设工程。

3. 土木工程设计师必须在电气、机械、供暖和通风的设计要求上与他人妥协。

Ⅴ. Answer the following questions briefly.

1. What are the four distinct phases of all civil engineering projects according to time order?
2. In the first phase, to make the most important decisions, what questions must be answered?
3. There are a number of different forms of schedules that are used in construction, please give some examples and draw one of them. What are the general principles that different schedules should observe?
4. Constructing a civil engineering project needs many different techniques, present some examples and describe one of them in details.

Reading Materials

FIDIC Tendering Procedure

The International *Federation*(联合) of Consulting Engineers (FIDIC) published the first edition of *Tendering*(投标) Procedure in 1982. The first edition primarily addressed procedures which FIDIC recommended for the selection of tenders and the preparation and evaluation of tenders for civil engineering contracts. The document reflected the provisions of the then current (third edition) of the FIDIC Conditions of Contract (International) for Works of Civil Engineering Construction.

This document presents a systematic approach for tendering and awarding of contracts for international construction projects. It is intended to assist the employer/engineer to receive sound competitive tenders in accordance with the tender documents so that they can be quickly and efficiently assessed. At the same time, an effort has been made to provide the opportunity and *incentive*(刺激,鼓励) to contractors to respond easily to invitations to tender for projects which they are qualified to implement. It is hoped that the adoption of this procedure will minimize tendering costs and ensure that all tenders receive a fair and equal opportunity to submit their offers on a reasonable and comparable basis.

Prequalification of tenders

Prequalification documents should give information about the project, the tendering procedure and the prequalification procedure. They should also specify what data is required from contractors wishing to prequalify. The documents are prepared by the employer/engineer and will normally include the following: letter of invitation to prequalify; information about the prequalification procedure; project information; prequalification application.

The employer/engineer should publish a notice inviting interested contractors to apply for prequalification documents, stating that tender documents will be issued only to a limited number of companies/*joint ventures*(合资企业) selected by the employer/engineer as having the necessary qualifications to perform the work satisfactorily.

The notice should be published in appropriate newspapers and *journals*(期刊) to give sufficient publicity according to the particular circumstances of the project. The notice may also be issued to financing institution representatives, if relevant, and to government agencies responsible for foreign trade so that the international community receives timely notification of the proposed project and instructions on how to apply.

The notice should be reasonably brief and where *feasible*(可能) contain: ① name of the employer; ② name of the engineer; ③ location of the project; ④ description of the project and scope of work; ⑤ source of finance; ⑥ anticipated program (i. e. award of contract, completion and any other key dates); ⑦ planned dates for issue of tender documents and submission of tenders; ⑧ instructions for applying for prequalification documents; ⑨ date by which applications to prequalify must be submitted; ⑩ minimum qualification requirements and any particular aspects which could be of concern to *prospective*(预期的) tenders.

The period between the notice of invitation to prequalify and the latest date for the return of completed applications should not be less than four weeks.

The employer/engineer should evaluate the prequalification applications to identify those companies/joint ventures which they consider to be suitably qualified and experienced to undertake the project, if the resulting list, after those firms that were found unsuitable have been excluded, exceeds six potential tenders and there are no special regulations or conditions imposed on the employer, the selection procedure should be continued to eliminate the less well-qualified in order to arrive at no more than six.

When the list of selected tenders has been prepared, successful applicants should be notified and requested to confirm their intention to submit a tender. This should ensure, as far as possible, an adequate number of competitive tenders. If a potential tender wishes to drop out at this stage, the next best-placed should be invited and asked to confirm as above. Following this, all applicants should be notified of the list of selected tenders without giving explanation for the decisions.

Obtaining tenders

The tender documents prepared by the employer/engineer will normally include the following.

- Letters of invitation to tenders.
- Instructions to tenders.
- Tender form and *appendices*(附录).
- Conditions of contract (Parts I and Ⅱ) together with any requisite forms.
- *Specification*(详细说明书).
- Drawings.
- Bill of quantities or schedule of prices.
- Information date.
- List of additional information required from tenders.

Instructions to tenders should be prepared by the employer/engineer to meet the particular requirements of individual contracts. The purpose of the document is to convey information and instructions that will govern the preparation, submission and evaluation of tenders.

When determining the tender period, the employer/engineer must ensure that adequate time is available for tenders to prepare their tenders, taking into account the size, complexity and location of the project in question.

Tenders should be notified of the number of copies of their tender that are required, stipulating that one set of the documents should be, clearly marked "Original Tender" and the others (which should be photocopies) marked "Copy" and that, in the event of discrepancy, the "Original Tender" shall take *precedence*(优先).

The instructions to tenders should state that the employer does not bind himself to award a contract to any of the tenders.

Tenders should be advised of the source of finance and related conditions. When tenders are required to provide finance, they should be instructed to provide information as to source of finance and the conditions which will apply.

Specific instructions should be given concerning the currencies to be used in the preparation of the tender. Tenders should also be advised in which currency/currencies payments will be effected.

The requirements for a tender security, if any, will be determined by the circumstances of each project. If a tender security is required, a form should be included in the tender documents. The amount and currency/currencies of the security should be stated. In all cases the surety or sureties must be satisfactory to the employer, if a tender security has been requested; any tender which has not been so secured will specify.

Specification

The specification will define the scope and technical requirements of the contract, including all requirements for training and the transfer of technology. The quality of materials and the standards of workmanship to be provided by the contractor must be clearly described, together with requirements for quality assurance to be performed by the contractor and the required safety, health and environmental measures to be observed during the executions of the works, the extent, if any, to which the contractor will be responsible for the design of the *permanent*(永久的) works should also be specified. Details should be included of samples to be provided and tests to be carried out by the contractor during the course of the contract. Any limitations on the contractor's freedom of choice in the order, timing or methods of executing the work or sections of the works would be clearly set out and any restrictions in his use of the site of the work, such as interface requirements with other parts of the work, or provision of access or space for other contractors, should be given.

The specification shall promote the broadest possible competition and as far as possible follow international standards such as those issued by ISO.

Drawings

The drawings included in the tender documents should provide tenders with sufficient details to enable them, in conjunction with the specification and the bill of quantities, to make an accurate assessment of the nature and scope of the works included in the contract. The drawings should be listed in the specification.

Evaluation of tenders

Following the opening, tenders should be checked by the employer/engineer to establish that they are arithmetically correct, are responsive without errors and omissions and consistent with the invitation to tender.

The evaluation of tenders can generally be considered to have three components. The components may include: technical evaluation; financial evaluation; general contractual and administrative evaluation.

Award of contract

The employer will normally seek to award the contract to the tender submitting the lowest evaluated responsive tender. The award must be made during the period of tender *validity*(有效性，合法性) or any extension thereto accepted by the tenders. The contractor should normally be required to sign a contract agreement with the employer. The employer/engineer should prepare the contract agreement which should include the following documents.

- Letter of acceptance and *memorandum*(备忘录) of understanding.
- Letter of Intent(if applicable).
- The tender.
- Conditions of contract.
- Specifications.

- Drawings.
- Bill of quantities.
- and such other documents that are intended to form the contract.

科技英语写作技巧(12)——科技论文的组成

科技论文的写作有一定的模式,常见的在正式期刊上发表的文章一般包括如下内容:①论文标题;②作者姓名及单位标属;③摘要;④关键词;⑤引言;⑥正文;⑦结论;⑧附录;⑨致谢;⑩参考文献等。

下面介绍科技论文的主体及所涉及的内容。

1. 引言(introduction)。

引言用来简要介绍文章主题、写作背景、写作目的、相关课题的现状及存在的问题、文章所要重点说明的问题及各节标题等。多数论文的引言比较短,通常约占科技杂志的半页左右。

2. 正文(body)。

正文是一篇论文的关键部分,所要论述的内容包括一些设想、实验装置、实验情况、获得的数据、证实的理论或新的设计方法等。作者可通过定义、描写、说明、举例、实验、论证、对比、分析、推理等不同的研究过程来说明在引言中所提出的主题。

3. 结论(conclusions)。

结论就是作者通过正文的论述而形成的总的观点,是一种概括性很强的论断,是整篇论文的结尾。结论必须完整、鲜明、准确、严谨。结论之后作者一般会提出建议,即根据所得结论对后续的研究工作提出意见或方案。

结论部分的标题通常用 conclusion(s)或 concluding remarks;有时结论中会罗列研究结果,这时标题用 results;有时结论中会就某一方面的问题加以讨论,标题则用 discussion,此时篇幅较长。

4. 附录(appendix)。

并不是所有论文都要有附录,它主要是对论文中提及的某个定理加以证明,对某项内容进行推导或针对某一点作进一步的说明等。有的论文将这一部分放在致谢后面。

5. 致谢(acknowledgement)。

致谢也不是每篇论文必有的项目。但如果论文的完成得到了他人的帮助,则必须要提出感谢。所以这一部分作者用来表达对某人或某单位的谢意。

6. 参考文献(references)。

英语科技论文的参考文献可以列在每一页的脚注中,也可以放在正文之后。文后参考文献须另起一页,并且按照在论文中出现的顺序或按姓氏的字母顺序进行排列。

Lesson 26

CIVIL ENGINEERING CONTRACTS

A simple contract consists of an agreement entered into by two or more parties, whereby one of the parties undertakes to do something in return for something to be undertaken by the other. ① A contract has been defined as an agreement which directly creates and contemplates an obligation. The word is derived from the Latin "contractum", meaning drawn together.

We all enter into contracts almost every day for the supply of goods, transportation and similar services, and in all these instances we are quite willing to pay for the services we receive. Our needs in these cases are comparatively simple and we do not need to enter into lengthy or complicated negotiations and no written contract is normally executed. Nevertheless, each party to the contract has agreed to do something, and is liable for breach of contract if he fails to perform his part of the agreement.

In general, English law requires no special formalities in making contract but, for various reasons, some contracts must be made in a particular form to be enforceable and, if they are not made in that special way, then they will be ineffective. Notable among these contracts are contracts for the sale and disposal of land, and "land", for this purpose, includes anything built on the land, as, for example, roads, bridges and other structures.

It is sufficient in order to create a legally binding contract, if the parties express their agreement and intention to enter into such a contract. ② If, however, there is no written agreement and a dispute arises in respect of the contract, then the Court that decides the dispute will need to ascertain the terms of the contract from the evidence given by the parties, before it can make a decision on the matters in dispute.

On the other hand if the contract terms are set out in writing in document, which the parties subsequently sign, then both parties are bound by these terms even if they do not read them. Once a person had signed a document, he is assumed to have read and approved its contents, and will not be able to argue that the document fails to set out correctly the obligations which he actually agreed to perform. Thus by setting down the terms of a contract in writing one secures the double advantage of affording evidence and avoiding disputes.

The law relating to contracts imposes on each party to the contract a legal obligation to perform or observe the terms of the contract, and gives to the other party the right to enforce the fulfillment of these terms or to claim "damages" in respect of the loss sustained in consequence of the breach of contract.

Most contracts entered into between civil engineering contractors and their employers are of the type known as "entire" contracts. These are contracts in which the agreement is for special works to be undertaken by the contractor and no payment is due until the work is complete.

In an entire contract, where the employer agrees to pay a certain sum in return for civil engineering work, which is to be executed by the contractor, the contractor is not entitled to any payment if he abandons the work prior to completion, and will be liable in damages for breach of contract. Where the work is abandoned at the request of employer, or results from circumstances that were clearly foreseen when the contract was entered into and provided for in its terms, the contractor will be paid as much as he has earned.

It is, accordingly, in the employer's interest that all contracts for civil engineering work should be entire contracts to avoid the possibility of work being abandoned prior to completion. However, contractors are usually unwilling to enter into any contracts, other than the very smallest, unless provision is made for interim payments to them as the work proceeds. For this reason the standard form of civil engineering contract provides for the issue of interim certificates at various stages of the works.

It is customary for the contract further to provide that a prescribed proportion of the sum due to the contractor on the issue of a certificate shall be withheld. This sum is known as "retention money" and serves to insure the employer against any defects that may arise in the work. The contract does, however, remain an entire contract, and the contractor is not entitled to receive payment in full until the work is satisfactorily completed, the maintenance period expired and the maintenance certificate issued.

The work must be completed to the satisfaction of the employer, or his representative, and does not give the employer the right to demand an unusually high standard of quality throughout the works, in the absence of a prior express agreement. Otherwise the employer might be able to postpone indefinitely his liability to pay for the work. The employer is normally only entitled to expect a standard of work that would be regarded as reasonable by competent persons with considerable experience in the class of work covered by the particular contract. The detailed requirements of the specification will have a considerable bearing on these matters.

The employer or promoter of civil engineering works normally determines the conditions of contract, which define the obligations and performances to which the contractor will be subject. He often selects the contractor for the project by some form of competitive tendering, and any contractor who submits a successful tender and subsequently enters into a contract is deemed in law to have voluntarily accepted the conditions of contract adopted by the promoter.

The obligations that a contractor accepts when he submits a tender are determined by the form of invitation to a tender. In most cases the tender may be withdrawn at anytime until it has been accepted and may, even then, be withdrawn if the acceptance is stated by the promoter to be subject to formal contract as is often the case.

The employer does not usually bind himself to accept the lowest or indeed any tender and this is often stated in the contract. A tender is, however, normally required to be a definite offer and acceptance of it gives rise legally to a binding contract.

A variety of contractual arrangements are available and engineers will often need to carefully select the form of contract which is best suited for the particular project. The employer is entitled to know the reasoning underlying the engineer's choice of contract.

Types of contracts are virtually classified by their payment system: ①price-based: lump sum and admeasurements (prices or rates are submitted by the contractor in his tender); and ②cost-based: cost-reimbursable and target cost (the actual costs incurred by the contractor are reimbursed, together with a fee for overheads and profit).

Words and Expressions

contract	[ˈkɔntrækt]	*n.* 合同,契约
obligation	[ˌɔbliˈgeiʃən]	*n.* 义务,责任
breach	[briːtʃ]	*n.* & *v.* 破坏,违反
promoter	[prəˈməutə]	*n.* 发包者
tender	[ˈtendə]	*n.* 标书
		v. 招投标
overhead	[ˈəuvəhed]	*n.* 管理费
profit	[ˈprɔfit]	*n.* 利润
(be) liable for		对……负责的
breach of contract		违约
binding contract		有(法律)约束力的合同
contract term		合同条款
claim "damages"		索赔
interim payment		进度款
retention money		保留(滞留)金

Notes

① A simple contract consists of an agreement entered into by two or more parties, whereby one of the parties undertakes to do something in return for something to be undertaken by the other.

whereby=by which,意为"借此",引出一个非限定性定语从句。全句意思为:一份简单合同由两方或多方参与的协议组成,借此(协议)一方承担某项任务去换取他方所承担的责任。

② It is sufficient in order to create a legally binding contract, if the parties express their agreement and intention to enter into such a contract.

It is sufficient in order to create…句中,it 为形式主语,不定式 to create a legally binding contract 为真实主语,介词短语 in order 的含义是"符合要求,妥当",充当状语,修饰 sufficient。

Exercises

Ⅰ. Translate the following words into Chinese.

1. enter into lengthy or complicated negotiations ________
2. legally binding contract ________
3. perform or observe the terms of the contract ________
4. pay a certain sum in return for civil engineering work ________
5. the satisfaction of the employer ________
6. the reasoning underlying the engineer's choice of contract ________

Ⅱ. Translate the following words into English.

1. 两方当事人订立的协议________
2. 订立合约________
3. 自愿接受合同条款________
4. 竞争性招标________

Ⅲ. Translate the following sentences into Chinese.

1. The contractor is not entitled to receive payment in full until the work is satisfactorily completed, the maintenance period expired and the maintenance certificate issued.
2. The employer does not usually bind himself to accept the lowest or indeed any tender and this is often stated in the contract.

Ⅳ. Translate the following sentences into English.

1. 最终,由法院根据合同条款和各方证词来裁决争议。
2. 若承包商不履行维修期的合同条款,业主有权拒付保留金。

Ⅴ. Answer the following questions briefly.

1. Why does the author say that everybody enter into contracts every day?

2. How do you understand "entire" contracts?
3. Can you tell the differences between price-based and cost-based contracts?

Reading Materials

Other Forms of Contract Project

The *competitive bidding system*(招标制)of construction contracts is the norm for public works in Japan. This system is also widely used in contracts for private projects. Besides this type of contract, a variety of others—such as turn-key contracts and CM contracts—have been used domestically in other advanced countries and in international projects for the Middle East and Southeast Asia.

These forms of contract have been *devised*(设计) to suit the social and economic background peculiar to each country. In addition, various management theories and methods of application developed in the West since the 1970s have come into practical use in response to the needs of construction projects operated as businesses.

In this section some forms of contracts are described, namely the design-build or design-construction contract, the turn-key contract, the CM contract, the BOT contract, the partnering contract, and the contracts with a Value Engineering clause.

Design-build or design-construction contract

This is a form of contract in which the contractor takes on responsibility for both design and construction, based on plans drawn up by the client. The merit of this type of contract is that project *optimization*(优化) can be applied in a consistent manner, because with a single organizational structure, the design can easily take into account construction procedures and feedback from actual construction work. At the same time, the contractor has to take on responsibility for the whole project, and must prove itself through a well-balanced approach to design and construction. In most cases, a cost-plus-fee contract or a lump sum contract that includes both design and construction costs is adopted. Contracts of this form are often adopted when the client has no in-house design and engineering departments, and when *subcontracting*(分包) the design work alone is *inappropriate*(不适当的，不相称的).

In the West, large construction companies with their own design departments are known as design constructors, and they often participate in public works such as road construction under contracts of this form. In America, the effects of design-build or design-construction contracts are seen as follows.

① Comprehensive approach reduces construction cost.

② Technological *prowess*(造诣) is stimulated.

③ Fewer *disputes*(争议,纠纷) and lawsuits.

However, there is also fairly strong opposition, on the grounds that design-build or design-construction contracts infringe traditional codes of practice and principle, since design and engineering were traditionally independent functions.

Turn-key contracts

This is a form of contract in which the contractor undertakes every aspect of the project under contract from the client: surveying, project planning, design, construction, and test operation. The term "turn-key" derives from the final handing over of a key by the contractor, which the client then simply turns to start the facility. This is therefore a comprehensive contract, and may also be known as an all-in contract. It first came to the fore in big projects, such as petrochemical plants and nuclear power stations, during the late 1960s in the United States. In many cases, payment is on a cost-plus-fee basis because such big projects are inherently risky.

The procedure for making a contract is as follows. The client first prepares documents stating the requirements of the facilities to be constructed and either selects the best proposal from those submitted by multiple bidders, or *designates*(指定) a specific contractor from the beginning and enters into a contract when negotiations begin. For the client, the system has a number of merits, including clarity of responsibility, and reduction of the time taken to prepare for bidding and carrying out examinations, because a single contractor handles all project phases.

In some cases, such comprehensive contracts include not only civil engineering and building work, but also the procurement and installation of equipment and systems. In this case, the contractor must display a thorough knowledge and have suitable capabilities in the equipment field. A frequently used term is "full-turn-key contract", meaning a contract which even includes the training of operators on top of the standard turn-key contract. The term "package deal contract" as often used in the United States has the same meaning as turn-key contract, while an extended turn-key contract is one in which the contractor also obtains financing for the project. In Germany, the respective terms are general *turnkey*(交钥匙) contractor of project developer.

Construction management contract

A construction management contract is a form of contract in which the client commissions a specific organizational structure (a construction manager) for the management of a project. The construction manager has responsibilities which include all project work: planning, design, procurement, order placement, construction, and handing over. However, in principle, the manager does not receive a contract for the design, construction, and other work. Rather, the construction manager acts as an agent for the client, a construction specialist, and receives actual costs plus fees in exchange for his services; this is thus different from a turn-key contract.

This is another kind of contract developed in the United States in the early 1970s and is now widely used. In the UK, this contract form is called a management contract. In Germany, it is called a project controller contract. The construction manager or project controller in such an arrangement must have the following capabilities.

① The cognitive power to continually improve and integrate the project with the client's needs, from the client's point of view.

② The capability of optimizing design while taking into account various project elements such as cost, schedule, quality, maintenance, and management. All in close cooperate with a designer.

③ The competence and experience needed to select capable subcontractors and quality construction materials, and to make the most of available construction planning and management techniques such as QC, OR, CPM, and computer systems.

The advantage for the client is that by choosing a construction manager with these skills, the best possible project execution will be achieved. It is a form of contract suited to complex, large-scale projects which must be completed within a limited construction period, and where orders for phased work can be placed.

As this type of contract developed in the United States, a variation known as "construction management at risk", in which a contract is offered for part of the construction work, depending on the needs of the project, was also conceived of. This has also been applied in many actual projects.

BOT contract

This is an abbreviation for the build, operate, and transfer system of contracting. In addition to the responsibilities of the turn-key contract, this type of contract includes fund raising as the project advances, operating the completed facilities for a predetermined period of time, and then a subsequent transfer to the client. Thus, it far exceeds the scope of a simple construction contract.

This form of contract was originally conceived by the late Turkish Prime Minister Ozal, and has been used in many debtor and developing countries. Lately, it has found favor in public works in advanced countries as a way to introduce the aggressiveness of private enterprise into public operations. Where a public project is likely to be profitable, such as in the case of toll roads, urban railways, and power plants, a BOT contractor raises the funds, undertakes all project work including design, engineering, and construction, and then operates the facilities for 10 to 30 years to recover the initial investment before finally handing the facilities over to the client.

The key question in this type of contract is how the client guarantees the contractor's operating income. The contractor must find freedom from risk through close cooperation with financial institutions and operation specialists. Usually, the best offer is selected from a number of BOT proposals submitted to the client and an order is then placed.

Partnering agreement

This is a new form of agreement or system, adopted within normal construction contracts or design-build contracts, in which the client and contractor together form a project team based on mutual confidence and then work together to manage the project to a successful conclusion, yielding a profit for both parties. It has been in use since the early 1980s in the United States, particularly in private projects. Although it has been used for military projects, examples in public works are very few.

The relationship between the two parties is called a partnership or alliance. One agreement is usually valid for a number of years, but agreements for shorter periods or those for a single project are also possible. The agreement typically covers planning, design, engineering, procurement, and construction supervision. Payment provision is made based on a cost-plus-fee basis.

This type of agreement was originally conceived with the aim of avoiding contractual confrontations and disputes. Since the formation of certain types of partnership has been limited in the United States over the past few years, the effectiveness and problems of this new system will be revealed in future studies.

Contract with a value engineering clause

Value Engineering (VE) is a management technique conceived and put into practice originally in the United States. It aims at actively adopting means of reducing costs without reducing *functionality*(功能), and making use of the many available ways of achieving the objectives. At the same time, it looks for ways of increasing functionality at equivalent cost. The technique is now internationally used.

The field of building construction provides particularly good opportunities for taking advantage of VE at all stages, from design through construction. Essentially, VE is an organizational activity which develops optimum improvement plans and translates them into action by taking into account all aspects of the project: quality, cost, schedule, safety, management and operation according to the client's original plans, construction methods, and technologies used. This activity can be carried out by the client and is a natural part of designer and consultant work. However, in most cases this system is proposed by the contractor.

The client benefits in a number of ways: construction costs can be reduced without compromising functionality, proprietary construction technologies and expertise can be used effectively, and up-to-date technical data can be accumulated. The benefit for the contractor is that remuneration is received in accordance with efforts to economize, such as by improving site productivity and reducing the construction period. Further, it encourages a competitive attitude to technology.

Then collecting and choosing from VE proposals, the following types of operation may opt for contracts with a VE clause.

① Bidding with a VE proposal. A tender is submitted, including the cost of improvements to the client's original plans.

② Contracting with a VE proposal. The contract price is adjusted (with remuneration), in accordance with the successful bidder's improvements to the original plans.

③ Placing orders with a VE promotion clause. After a contract has been concluded based on the client's original plans, improvements are submitted by the successful bidder and remuneration is paid if the plans are adopted.

There has been a rise of contracts including a VE clause in the West and in international projects. In Germany, in particular, this method has a long tradition, but without the term VE. Within the past 50 years, hardly any public engineering project such as bridges, tunnels, or foundations have been executed according to the client's original design. Rather, modified and optimized designs and methods proposed by the successful VE bidder were used.

科技英语写作技巧(13)——论文摘要(Ⅰ)

为了便于学术之间的交流，论文摘要常常被专业期刊文献杂志编入索引资料或文献刊物。此外，参加学术会议，一般要首先提供论文摘要，以期通过会议论文审议组的审定，取得参加会议的资格，所以学会写英文摘要是至关重要的。

1. 摘要的英文术语及概念。

摘要也称为内容提要，其英文术语为 abstract。它是对原文内容要点的总结，并以高度浓缩与概括的形式把原文的主要内容表达出来。通常在科技论文中都必须附有摘要，其位置应放在论文的正文之前。

根据联合国教科文组织的规定："全世界公开发表的科技论文，不管用何种文字写成，都必须附有一篇简练的英文摘要。"既然这个规定对摘要提出了这样一个要求，那么摘要的内容简练到何种程度才算是达到要求呢？学术论文摘要的长短一般约定为正文字数的 2%～3%。国际标准化组织建议不少于 250 个词，最多不超过 500 个词。摘要的文字既要精练到符合这个规定，而且还应该将论文的要点囊括其中。

2. 摘要的内容要求。

论文摘要一般包括以下四部分内容。

(1)研究目的(What do you want to do)。

(2)研究过程和方法(How do you want to do it)。

(3)结果、结论和建议(What results you can get and what conclusions you can draw)。

(4)创新之处(What is the originality in your paper)。

在撰写摘要的过程中,对所掌握的资料应进行精心筛选,不属于上述"四部分"的内容不必写入摘要。对属于上述"四部分"的内容,也应适当取舍,做到简明扼要,不能包罗万象。比如"目的",在多数标题中就已初步阐明,若无更深一层的目的,摘要完全不必重复叙述;再如"方法",有些在国外可能早已成为常规的方法,在撰写英文摘要时就可仅写出方法名称,而不必一一描述其操作步骤。

3. 摘要的文体要求。

(1)摘要应遵循准确、简明、清楚的原则。

准确是指内容上要忠实于原文。其内容是标题的扩充,是全文的高度概括。所以在摘要中不能加入自己的评论。简明是摘要文体要求中的一个重要内容,在摘要中应使用正规英语、标准术语,避免使用缩写词汇,也不能使用图表和表格。清楚是指使用简洁的正规英语将文章的论题、论点、实验方法、实验结果用有限的字数表达出来,要做到不遗漏、不重复,不使用带有感情色彩和意义不确定的词汇,也不使用祈使句和感叹句,要避免把各种时态混在一起使用。

掌握一定的遣词造句技巧的目的是便于简单、准确地表达作者的观点,减少读者的误解。但是,英文有其自身的特点,最主要的是中译英时往往造成所占篇幅较长,同样内容的一段文字,若用英文来描述,其占用的版面可能比中文多一倍。因此,撰写英文摘要更应注意简洁明了,力争用最短的篇幅提供最主要的信息。

(2)时态和语态。

选择适当的时态和语态,是使摘要符合英文语法修辞规则的前提。通常情况下,摘要中谓语动词的时态和语态并非千篇一律,而应根据具体内容而有所变化,否则容易造成理解上的混乱。但这种变化又并非无章可循,其中存在如下规律。

①叙述研究过程,多采用一般过去时。

②在采用一般过去时叙述研究过程时,若提及在此过程之前发生的事,宜采用过去完成时。

③说明某课题现已取得的成果,宜采用现在完成时。

④摘要开头表示本文所"报告"或"描述"的内容,以及摘要结尾表示作者所"认为"的观点和"建议"的做法时,可采用一般现在时。

⑤英文摘要仍以被动语态居多,这种语态可以在主语部分集中较多的信息,起到信息前置,语义鲜明突出的效果。主动语态也偶有出现,并有增长的趋势,认为主动语态表达的语句文字清晰、简洁明快,表现力强,动作的执行者和动作的承受者一目了然,往往给人一种干净利落的感觉。

(3)语法修饰。

① 力求简洁。

at a temperature of 250℃ to 300℃→at 250 to 300℃

at a high pressure of 2 kPa→at 2 kPa

has been found to increase→increased

from the experimental results, it can be concluded that→the results show that

② 能用名词作定语的不用动名词作定语,能用形容词作定语的不用名词作定语。

measuring accuracy→measurement accuracy

experiment results→experimental results

③ 可直接用名词或名词短语作定语的情况下,要少用 of 句型。

accuracy of measurement→measurement accuracy

structure of crystal→crystal structure

④ 可用动词的情况尽量避免用动词的名词形式。

Measurement of thickness of plastic sheet was made. →

Thickness of plastic sheet was measured.

Lesson 27

HIGHWAY ENGINEERING(1)

Transportation has always been one of the most important aspects of civil engineering. One of the greatest accomplishments of Roman engineers was the high way system that made rapid communication possible between Rome and the provinces of the empire. ① Modern highways are built according to the principles laid down in the eighteenth and early nineteenth centuries. Generally speaking, alignment, subgrade, and pavement are the most basic three parts of the modern highways.

Alignment

The alignment of a road includes horizontal alignment and vertical alignment.

The horizontal alignment of a road is shown on the plan view and is a series of straight lines called tangents connected by circular curves. In modern practice, it is common to interpose transition or spiral curves between tangents and circular curves. So the elements of the horizontal alignment are tangent, circular curve, and transition curve, as shown in following Fig. 27-1.

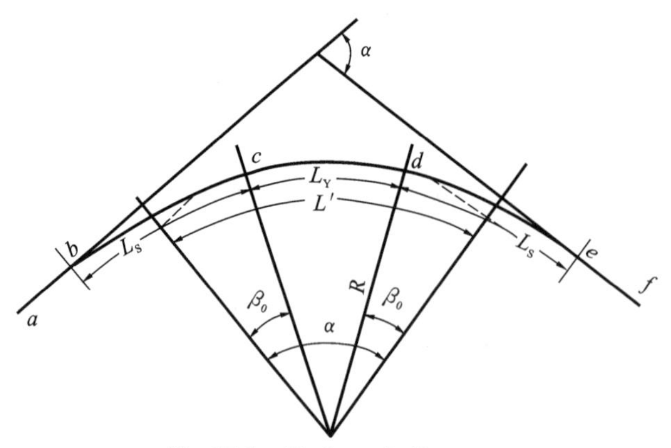

Fig. 27-1 Horizontal alignment

ab, *ef*—straight lines; *bc*, *de*—transition curves; *cd*—circular curve

The vertical alignment of a road is shown on the cutaway view. The elements of the vertical alignment are tangents, vertical curves. Usually there are two types of vertical curves: convex vertical curve and concave vertical curve, as shown in following Fig. 27-2.

Alignment must be consistent. Sudden changes from flat to sharp curves and long tangents followed by sharp curves must be avoided; otherwise, accident hazards will be created. Likewise, placing circular curves of different radii end to end (compound curves) or having a short tangent between two curves is poor practice unless suitable transitions between them are provided. ② Long, flat curves are preferable at all times, as they are pleasing in appearance and decrease possibility of future obsolescence. Long, flat curves should be used for small changes in direction, as short curves appear as "kink". Also horizontal and vertical alignment must be considered together, not separately. For example, a sharp horizontal curve beginning near a crest can create a serious accident hazard.

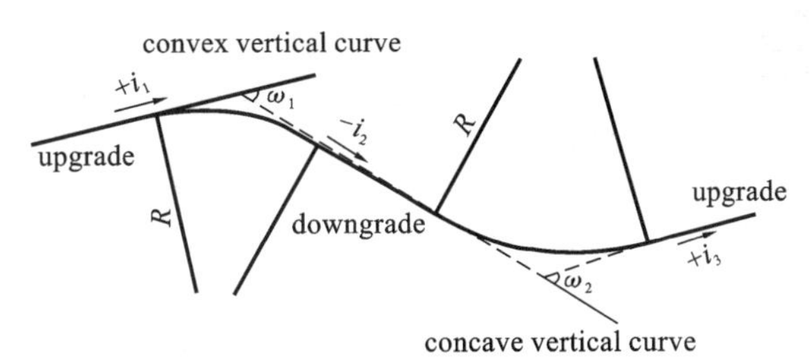

Fig. 27-2 Vertical alignment

A vehicle traveling in a curved path is subject to centrifugal force. This is balanced by an equal and opposite force developed through superelevation and side friction. From a highway design standpoint, both superelevation and side friction cannot exceed certain maximums, and these controls place limits on the sharpness of curves that can be used with a design speed. ③

When roads intersect or a railroad spans across a road, the road intersections are needed. The intersection can be at-grade intersection and grade-separated junction. The freeways only adopt the latter intersection. The typical at-grade crossings are cross roads, T intersection, and rotary intersection, as shown in following Fig. 27-3. The typical grade-separated junctions are clover-leaf interchange, diamond interchange, rotary interchange, and trumpet interchange, as shown in Fig. 27-4 and Fig. 27-5.

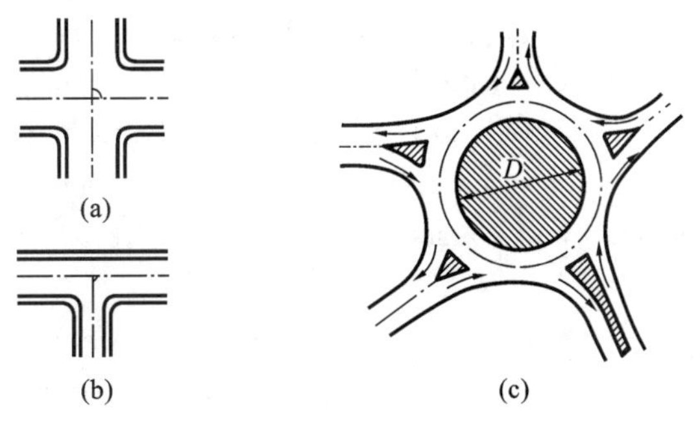

Fig. 27-3 At-grade intersection

(a) cross roads; (b) T intersection; (c) rotary intersection

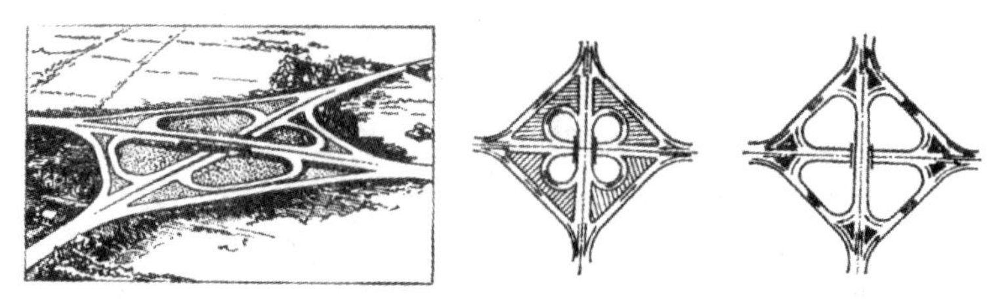

Fig. 27-4 Clover-leaf interchange

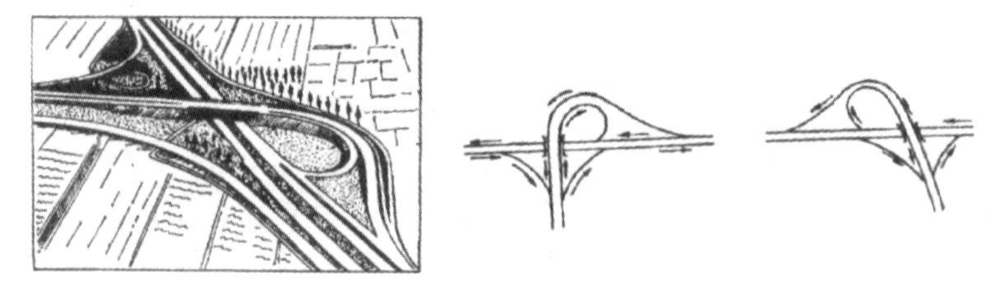

Fig. 27-5 Trumpet interchange

Highway Subgrade

The foundations under pavement of highways are subgrade, so highway subgrade (or basement soil) may be defined as the supporting structure on which pavement and its special under courses rest. ④ The purpose of a pavement is to provide a smooth surface over which vehicles may pass under all climate conditions. Because natural terrains are uneven, to get smooth road, cutting and filling are needed to change the natural terrain. Subgrades are classified into embankment, cutting, and part-cut part-fill subgrade, as shown in Fig. 27-6.

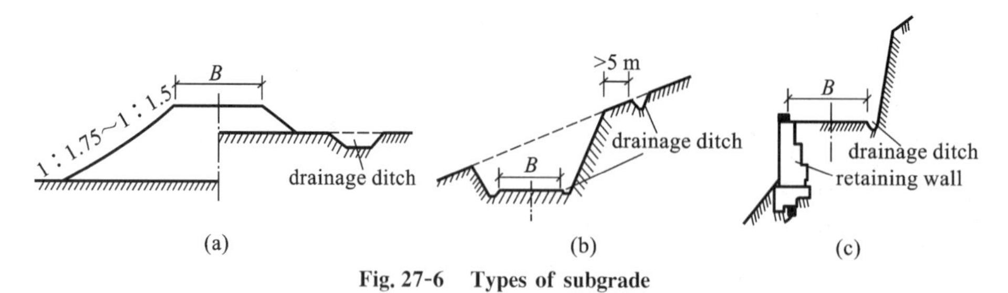

Fig. 27-6 Types of subgrade

(a) embankment; (b) cutting; (c) part-cut part-fill subgrade

In cut sections, the subgrade is the original soil lying below the special layers designated as base and subbase material. In fill sections, the subgrade is constructed over the native ground and consists of imported material from nearby roadway cuts or from borrow pits.

The performance of the pavement is affected by the characteristics of the subgrade. Desirable properties that the subgrade should possess include strength, drainage, ease of compaction, permanency of compaction, and permanency of strength.

Soils engineers agree that the properties of a soil mixture are influenced more by moisture than by any other cause. Soils that have ample strength and supporting power under one set of moisture conditions may be entirely unsatisfactory if the percentage of moisture changes. One difficulty with soils in highway subgrades is that they are subject to such moisture changes. That is the reason why drainage is an important consideration of subgrades. Since subgrades vary considerably, it is necessary to make a thorough study of the soils in place. Soil is a highly variable material; the interrelationship of soil texture, density, moisture content, and strength are complex, and behavior under repeated loads is difficult to evaluate. Because of the complexity of the problem, it is not possible to set down rules that will be suitable for all cases. For example, the supporting power of soils increases with density. A road constructed on a given soil may be entirely satisfactory if the soil is properly compacted. On the other hand, the road may fail if the soil is insufficiently compacted, particularly if the voids become filled with water.

Words and Expressions

subgrade	[ˈsʌbgreid]	*n.* 路基
transition	[trænˈziʃən, -ˈsiʒən]	*n.* 转变,转换,过渡
radii	[ˈreidiai]	radius 的复数形式
obsolescence	[ˌɔbsəˈlesəns]	*n.* (逐渐)废弃,退化
centrifugal	[senˈtrifjugəl]	*adj.* 离心的
superelevation	[ˈsjuːpəˌeliˈveiʃən]	*n.* (铁路或公路的)超高
cutting	[ˈkʌtiŋ]	*n.* 路堑
base	[beis]	*n.* 基层
subbase	[ˈsʌbˌbeis]	*n.* 底基层
pit	[pit]	*n.* 深坑,深渊
moisture	[ˈmɔistʃə]	*n.* 湿度
compact	[ˈkɔmpækt]	*v.* 压紧,压实
plan view		平面图,俯视图
spiral curve		螺线
transition curve		缓和曲线
cutaway view		剖面图
convex vertical curve		凸形竖曲线
concave vertical curve		凹形竖曲线
side friction		侧向摩擦力
at-grade intersection		平面交叉
grade-separated junction		立体交叉
clover-leaf interchange		苜蓿叶立体交叉
diamond interchange		菱形立体交叉

rotary interchange　　环形立体交叉
trumpet interchange　　喇叭形立体交叉
part-cut part-fill subgrade　　半挖半填式路基

Notes

① One of the greatest accomplishments of Roman engineers was the high way system that made rapid communication possible between Rome and the provinces of the empire.
句中 that 引导定语从句，made rapid communication possible，意思是：使快速通讯成为可能。

② Likewise, placing circular curves of different radii end to end (compound curves) or having a short tangent between two curves is poor practice unless suitable transitions between them are provided.
句中 placing circular curves of different radii end to end (compound curves) or having a short tangent between two curves 为两个并列的动名词短语做主语，意思是：把不同半径的圆曲线尾尾相连，或两条曲线之间插入短切线。unless suitable transitions between them are provided 为状语从句，意为：除非曲线之间有适当的缓和曲线。

③ From a highway design standpoint, both superelevation and side friction cannot exceed certain maximums, and these controls place limits on the sharpness of curves that can be used with a design speed.
句中 these controls 指代 the certain maximums of superelevation and side friction。that can be used with a design speed 为定语从句，修饰 sharpness of curves。全句的意思为：从公路设计的观点来讲，超高和侧向摩擦力都不能超过某个最大值，这些控制值设置了曲线锐度限值，曲线锐度限值可以与设计速度建立联系。

④ The foundations under pavement of highways are subgrade, so highway subgrade (or basement soil) may be defined as the supporting structure on which pavement and its special under courses rest.
句中 on which 引导定语从句修饰 supporting structure，on 也可置于句子末尾，形成固定短语 rest on，意思为：搁置路面及其特殊下层的结构。

Exercises

Ⅰ. Translate the following words into Chinese.

1. highway alignment ______________________
2. transition curve ______________________
3. clover-leaf interchange ______________________
4. centrifugal force ______________________
5. subgrade ______________________

Ⅱ. Translate the following words into English.

1. 路面__________
2. 平面交叉__________
3. 急弯，锐曲线__________
4. 设计速度__________
5. 路堑__________

Ⅲ. Translate the following sentences into Chinese.

1. The horizontal alignment of a road is shown on the plan view and is a series of straight lines called tangents connected by circular curves.
2. In cut sections, the subgrade is the original soil lying below the special layers designated as base and subbase material.
3. For example, a sharp horizontal curve beginning near a crest can create a serious accident hazard.
4. Desirable properties that the subgrade should possess include strength, drainage, ease of compaction, permanency of compaction, and permanency of strength.
5. Soil is a highly variable material; the interrelationship of soil texture, density, moisture content, and strength are complex, and behavior under repeated loads is difficult to evaluate.

Ⅳ. Translate the following sentences into English.

1. 目前在实际操作中，一般在切线和圆曲线之间插入缓和曲线和螺旋曲线。
2. 必须避免平地突然变化为急弯，以及很长的切线后紧跟着锐曲线，否则就会产生事故风险。
3. 因为自然地形不是平坦的，为了得到平坦的道路，需要挖方和填方来改变自然地形。
4. 在一定湿度下有足够强度与支撑能力的土体，如果湿度的百分比发生变化，土体可能不能完全令人满意。
5. 另一方面，如果土没有被充分压缩，特别是如果土体的孔隙充满了水，道路可能会失效。

Ⅴ. Answer the following questions briefly.

1. What are the main parts of the modern highways?
2. Please list the elements of horizontal alignment and vertical alignments of highways.
3. For a freeway, which road intersection can be adopted?
4. Why drainage is an important consideration of highway subgrade?

Reading Materials

Highway Engineering(2)

Highway pavement

To build the pavement, one layer material or multilayer different materials are paved on the subgrade to offer a level, safe traveling surface on which vehicles can pass. The function of the pavement structure is to distribute imposed wheel loads over a large area of the natural soil. If vehicles were to travel on the natural soil itself, shear failures would occur in the wheel path in most soils, and *ruts*(车辙) would form. The shear strength of the soil is usually not high enough to support the load.

Pavements are classified as "rigid" or "flexible", depending on how they distribute surface loads. The flexible pavement consists of a relatively thin *wearing surface*(磨损面层) built over a base course, and subbase course, and they rest upon the compacted subgrade, as shown in Fig. 27-7. The thickness of the *flexible pavement*(柔性路面) is meant to include all components of the pavement above the compacted subgrade. Thus, the subbase, base, and wearing surface are the structural components of the pavement. In contrast, *rigid pavements*(刚性路面) are made up of Portland cement concrete and may or may not have a base course between the pavement and the subgrade. In the case of the rigid pavement, the concrete, exclusive of the base, is referred to as the pavement. In some cases, Portland cement concrete is used as a base course for a flexible-type wearing surface. *Bituminous concrete pavement*(沥青混凝土路面), *bituminous penetration pavement*(沥青贯入式路面), and *clay-bound macadam pavement*(泥结碎石路面) are flexible pavement. Cement concrete pavement is rigid pavement.

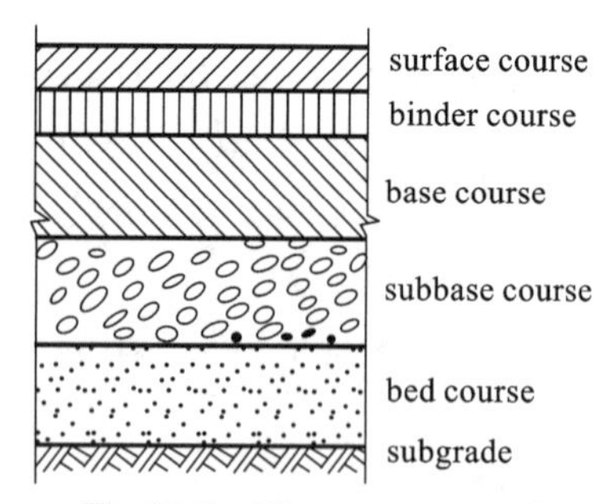

Fig. 27-7 The components of flexible pavement

The essential difference between the two types of pavements, flexible and rigid, is the manner in which they distribute the load over the subgrade. The rigid pavement, because of its rigidity and high modulus of elasticity, tends to distribute the load over a relatively wide area of soil; thus, a major portion of the structural capacity is supplied by the slab itself. The major factor considered in the design of rigid pavements is the structural strength of the concrete. For this reason, minor variations in subgrade strength have little influence upon the structural capacity of the pavement. The load-carrying capacity of flexible pavement is brought about by the load-distributing characteristics of the layered system. Flexible pavements consist of a series of layers, with the highest quality materials at or near the surface. Hence, the strength of a flexible pavement is a result of building up thick layers and thereby distributing the load over the subgrade, rather than by the bending action of a slab. The thickness design of the pavement is influenced by the strength of the subgrade.

Subbase courses for flexible pavements are generally made up of cheap, locally available materials, whereas the base courses are higher quality processed materials. In most cases the base course consists of crushed stone and, in some instances, may contain asphalt.

Pavement design consists of two broad categories-design of the paving mixture and structural design of the pavement components. The structural design of pavements is basically different from that of bridges and buildings, in that at the present time no rational method of design is available. Most methods of design are either empirical or semiempirical in nature and are based upon correlations with field performance.

Total load influences the thickness requirements for flexible pavements. *Tire pressures*(轮压) do not control the thickness of pavement to any great extent, but they influence the required quality of surface and base appreciably. Total load also influences the thickness of rigid pavements; tire pressures affect the thickness of pavement to a lesser extent.

Cross sections of typical highways of modern design are shown in Fig. 27-8 below. Dimension for each element are based on careful analysis of the *volume*(流量), character, and speed of traffic and of the characteristics of motor vehicles.

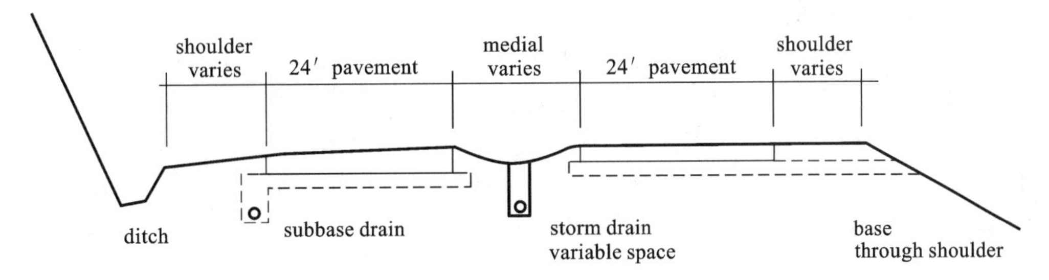

Fig. 27-8 Cross section of typical highway

科技英语写作技巧(14)——论文摘要(Ⅱ)

1. 摘要的写作技巧与注意事项。

(1)中英文摘要的一致性。

这里主要是指内容方面的一致性。目前对一致性这个问题的认识存在两个误区。

① 认为两个摘要的内容“差不多就行”,因此在英文摘要中随意删去中文摘要的重点内容,或随意增补中文摘要所未提及的内容,这样很容易造成英文摘要重心转移,甚至偏离主题。

② 认为英文摘要是中文摘要的硬性对译,对中文摘要中的每一个字都不敢遗漏,这往往使英文摘要用词累赘、重复,显得拖沓、冗长。英文摘要应严格、全面地表达中文摘要的内容,不能随意增删,但这并不意味着一个字也不能改动,具体撰写方式应遵循英文语法修辞规则,符合英文专业术语规范,并照顾到英文的表达习惯。

(2)精炼。

根据摘要的写作要求与文体特点,它应是论文的高度浓缩。所以摘要包括的内容是论文中课题研究的目的、研究过程(含实验方法与结果)、结论以及对下一步工作的启发。

精炼是摘要文体特征所要求的。如果当前的研究是在他人研究工作的基础上发展出来的,则必须说明研究的根据,也需要提到别人的研究。遇到这种情况,则需要用 On the basis of…'s research/conclusion/theory/methods of…或者用 According to…'s theory 等表达方式即可。不要使用繁杂的表达方式,如:After an extended series of trials over a period of several months following…theory/conclusion…

(3)具体翔实。

摘要的每个论点都要具体鲜明,直接讲论文“说明什么”,不要使用一些概念模糊的词汇,如果写作“not all the primary minerals are detrital; some are formed in place”,因为其中的 some 使人费解,不知是指哪一些,造成了意义上的模糊。

(4)完整。

完整实际上也是准确、清楚。摘要的内容要完整,这是因为读者是利用摘要或索引进行研究工作的,因此在摘要中一定要说明论文的主要内容,不要使用论文的导言或者某张插图来代替摘要中的内容。

(5)避免句子结构单调呆板。

在一篇摘要里(当然整篇论文也一样),短句和长句,主动语态和被动语态,名词短语、动词短语,以及各类短语最好交替使用。以使句型多样化。下面这个摘要句型都一样,缺乏技巧:The influence of … are discussed. Some design considerations are given, A practical construction … is shown. The method with high precision … is presented. The measured stability data … are also given.

(6)摘要的位置。

摘要通常是放在正文的前面。但是它的完成是在论文写作之后,而不是论文写作之前。这是因为论文在写作过程中要经过反复修改,内容在修改中可能进行调整。所以,在论文完成之后再写摘要是比较恰当的。

2. 摘要写作中一些有用的表达。

(1)以下表达经常用于摘要的第一部分。

① The author tells about … /The writer describes NP/The paper explores/looks at/deals with/ describes/ refers to…/According to the article/report/thesis/

paper … 等。

② The purpose of this paper is to ＋v. /A brief presentation of … is given / The concept of … is used to determine … /An investigation was designed to ＋v. 等。

③ 表示研究范围的表达：(to be) carried out/performed/made/conducted/studied 或(to be) investigated/described。

(2)以下表达经常用于摘要的第二部分。

① It is found that … /The author concludes that＋S＋v. /The paper concludes … 等。

② It is seen from Fig. 4 that … /Fig. 4 indicates/shows/provides that … /The results are given in Fig. 4 /The methods are given in Fig. 4 等。

(3)以下表达经常用于摘要的最后部分。

① It is suggested that … /It is recommended that … /It is recommended＋that ＋S＋v. 等。

② The paper suggests …

③ The results indicate (show/demonstrate/reveal/suggest) that …

(4)以下表达经常用于摘要中表示研究范围。

① (to be) carried out/performed/made/conducted/studied/investigated/described。

② deal with/elucidated/given/presented/developed/employed/derived/prepared/synthesized/monitored/determined/measured/observed/recorded/examined/characterized/identified/tested/calculated/proposed/used to study/used to establish/evaluated/discussed。

(5)表示转折与总结的词与词组。

as has been noted/to sum up/in summary/in brief short/in a word 等。

(6)表示试验或者研究采取的手段的词与词组。

by means of/by using/by the use of/using … as …等。

(7)表示试验或者其他研究结果和结论，可以用 That 从句。

The results indicated that …

The results show that …

The results demonstrated that …

The results reveal that …

The results are presented in the form of …

The author concludes/suggests that …

(8)(不)相符的表示方法。

… (to be) in good agreement with

… (to be) found to agree well with …

… (to be) consistent with …

… (to be) essentially identical with …

… (to be) found to coincide essentially with …
… (to be) closely analogous to …
… (to be) similar to …
Good agreement (to be) found between …
… (to be) contrary to …
… (to be) in contrast with …

Lesson 28

TIDES

The tide is the periodic rise and fall of the ocean waters produced by the attraction of the moon and sun. The movement is most noticeable on shores which shelve gradually and expose a wide expanse of beach between high and low water tide levels. Generally, the average interval between successive high tides is 12 hours 25 minutes, half the time between successive passages of the moon across a given meridian. The moon exerts a greater influence on the tides than the sun. This influence varies directly as the mass and inversely as the cube of the distance, and therefore the ratio is about 7 : 3. ①

The highest tides which occur at intervals of half a lunar month are called spring tides. They occur at or near the time when the moon is new or full, i. e. when the sun, moon, and earth fall in line, and the tide-generating forces of the moon and sun are additive. When the lines connecting the earth with the sun and the moon form a right angle, i. e. , when the moon is in her quarters, then the actions of the moon and sun are subtractive.

Owing to the retardation of the tidal wave in the ocean by frictional forces, as the earth revolves daily around its axis, and as the tide tends to follow the direction of the moon, the highest tide for each location is not coincident with conjunction and opposition, but occurs at some constant time after new and full moon. ② This interval is known as the age of the tide, which may amount to as many as two and one-half days.

Tides which occur twice each lunar day are called semidiurnal tides, and since the lunar day, or time it takes the moon to make a complete revolution around the earth, is about 50 minutes longer than the solar day, the corresponding high tide on successive days is about 50 minutes later. In some places, such as Pensacola, Florida, only one high tide a day occurs, and the tides are then called diurnal tides. If one of the two daily high tides is incomplete, i. e. , if it does not reach the height of the previous tide, as at San Francisco, California, then the tides are referred to as mixed diurnal tides. There are other exceptional tidal phenomena. For instance at Southampton, England, there are four daily high waters, occurring in pairs, separated by a short interval. At Portsmouth, there are two sets of three tidal peaks per day.

In using the tide tables of the United States Coast and Geodetic Survey, it must be kept in the mind that they give the times and heights of high and low waters and not the times of the flood or slack waters. ③ For stations on the ocean coast there is usually but little difference between the time of high or low water and the beginning of ebb or flood current, but for places in narrow channels landlocked harbors or along tidal rivers, the time of slack water may differ by several hours from the time of high or low water. The predicted times of slack water and tidal current velocities are given in tidal current tables published by the United States Coast and Geodetic Survey, one for the Atlantic Coast of North America and the other for the Pacific Coast of North America and Asia.

The rise of the tide is referred to some established datum of the charts, which varies in different parts of the world. ④ The British Admiralty charts use the level of mean low water springs; in the United States it is mean low water; in France and Spain it is the lowest low water. Mean high water is the average of the high water over a 19-year period, and mean low water is the average of the low waters over a 19-year period. Higher high water is the higher of the two high waters of any diurnal tidal day, and lower low water is the lower of the two low waters of any diurnal tidal day. Mean higher high water is the average height of the higher high waters over a 19-year period, and mean lower low water is the average height of the lower low waters over a 19-year period. Highest high water and lowest low water are the highest and lowest, respectively, of the spring tides of record. Mean range is the height of mean high water above mean low water. The mean of this height is generally referred to as mean sea level. Diurnal range is the difference in height between the mean higher high water and the mean lower low water.

Words and Expressions

tide	[taid]	*n*. 潮,潮汐
periodic	[ˌpiəriˈɔdik]	*adj*. 周期的,定期的
shelve	[ʃelv]	*v*. 置于架子上,搁置
meridian	[məˈridiən]	*n*. 子午线,正午
		adj. 子午线的,正午的
subtractive	[səbˈtræktiv]	*adj*. 减去的,负的,有负号的
retardation	[ˌriːtɑːˈdeiʃən]	*n*. 延迟
tidal	[ˈtaidl]	*adj*. 潮汐的,定时涨落的
frictional	[ˈfrikʃənəl]	*adj*. 摩擦的,摩擦力的
axis	[ˈæksis]	*n*. 轴
conjunction	[kənˈdʒʌŋkʃən]	*n*. 联合,关联,连接词
semidiurnal	[ˌsemidaiˈəːnəl]	*adj*. 12 小时的,半天的
corresponding	[ˌkɔrisˈpɔndiŋ]	*adj*. 相应的
phenomena	[fiˈnɔminə]	phenomenon 的复数形式

ebb	[eb]	*n.* 退，退潮，衰落 *v.* 潮退，衰退
high and low water tide levels		高、低潮汐水位
lunar month		[天]朔望月，太阴月
spring tide		大潮
tide-generating force		引潮力
right angle		直角
neap tide		小潮
new and full moon		朔日和望日
the age of the tide		潮候
semidiurnal tide		半日潮
the lunar day		太阴日
the solar day		太阳日
diurnal tide		全日潮
mixed diurnal tide		混合潮
in pairs		成对
slack water		平潮，平流，缓流
datum of the chart		海图基准面
the British Admiralty Chart		英国海军部海图
mean low water spring		大潮平均低水位
higher high water		较高高水位
diurnal range		日周期潮差
Pensacola, Florida		佛罗里达州的彭萨科拉
Southampton		南安普敦
Portsmouth		朴次茅斯

Notes

① This influence varies directly as the mass and inversely as the cube of the distance, and therefore the ratio is about 7:3.
(对潮汐的)影响与质量成正比例，与距离的立方成反比例，因此(月亮与太阳对潮汐影响的)比率为7:3。

② Owing to the retardation of the tidal wave in the ocean by frictional forces, as the earth revolves daily around its axis, and as the tide tends to follow the direction of the moon, the highest tide for each location is not coincident with conjunction and opposition, but occurs at some constant time after new and full moon.
由于海洋潮波受到摩阻力的阻滞作用、地球不停自转、潮水的涨落随着月球的方向而转移等原因，每一地区的最高潮位的发生时间不是与月球在朔、望日经过上中天的时间相合，而是比朔、望日时间迟一定的时间。

③ In using the tide tables of the United States Coast and Geodetic Survey, it must be kept in the mind that they give the times and heights of high and low waters and not the times of the flood or slack waters.
在使用美国海岸及大地测量局的潮汐表时应注意,表中所给出的是高、低潮位的时间和高度而不是涨潮或平潮的时间。

④ The rise of the tide is referred to some established datum of the charts, which varies in different parts of the world.
潮升是从海图基准面起算的,而世界各地的基准面不同。

Exercises

Ⅰ. Translate the following words into Chinese.

1. high and low water tide levels ________________
2. higher high water ________________
3. the age of the tide ________________
4. tide-generating forces ________________
5. mean low water springs ________________

Ⅱ. Translate the following words into English.

1. 平均潮差________________
2. 落潮流________________
3. 潮流速度________________
4. 潮升________________
5. 平均较低水位________________

Ⅲ. Translate the following sentences into Chinese.

1. The movement is most noticeable on shores which shelve gradually and expose a wide expanse of beach between high and low water tide levels.
2. Mean high water is the average of the high water over a 19-year period, and mean low water is the average of the low waters over a 19-year period.
3. If one of the two daily high tides is incomplete, i. e., if it does not reach the height of the previous tide, as at San Francisco, California, then the tides are referred to as mixed diurnal tides.

Ⅳ. Translate the following sentences into English.

1. 如果一天的两次潮汐中有一次高潮所达到的高度不及前一次的高潮,这一类型的潮汐称为混合潮。
2. 自第二次世界大战以来,人们通过气象、海洋和地理的资料来进行潮汐和波浪的预报工作。
3. 在浅水区成长的波,波高受到两个因素的限制,即底部摩擦和破碎。
4. 全日潮是指在有些地方一天只发生一次高潮的潮汐。
5. 如果港口被陆地环绕,它的平潮时间与高、低水位时间可能要相差几个小时。

Ⅴ. Answer the following questions briefly.

1. What is the major difference between diurnal tides and mixed diurnal tides?
2. What is the age of tide?
3. What is diurnal range?

Reading Materials

Waves

The behavior of water waves is one of the most intriguing and probably one of the least understood of nature's phenomena. Water waves may be caused by certain artificial disturbances such as moving vessels or explosions; or they may be caused by earthquakes, tides, or winds. It is the last which produces the waves in which engineers are most interested and which have the most influence on the design of marine structures. Generally, the tides, because of their slow rise and fall, have little effect on the formation of waves, except in the case of tidal bores. Tidal bores are a regular occurrence at certain locations and are a high crested single or solitary type of wave caused by the rush of flood tide up a river, as in the Amazon, or by the meeting of tides as in the Bay of Fundy.

Wave form and generation. Waves manifest themselves by curved undulations of the surface of the water occurring at periodic intervals, except for waves of translation and solitary waves or single waves of translation without any depression below still-water level.

Wave disturbance is felt to a considerable depth, and therefore the depth of water has an effect on the character of the wave. Deep-water waves are those which occur in water having a depth greater than one-half the wave length ($d>L/2$), at which depth the bottom does not have any significant influence on the motion of the water particles. Shallow-water waves are those which occur in water having a depth less than one-half the wave length ($d>L/2$), and the influence of the bottom changes the form of orbital motion from circular to elliptical or near-elliptical. Waves break when the forward velocity of the crest particles exceeds the velocity of propagation of the wave itself. In deep water this normally occurs when the wave height exceeds one-seventh of the wave length. When the wave reaches shallow water where the depth is equal to about one and one-quarter of its height it will usually break, although it may break in somewhat deeper water, depending upon the strength of the wind and the condition of the bottom.

An unbroken wave is a wave of *oscillation*(振荡), and even after breaking in deep water the wave will usually reform into an oscillatory wave of reduced height. It is only when it reaches shallow water and breaks without being able to reform that it becomes a wave of translation, a familiar sight in the form of breakers along the shore. The only pure wave of translation is the solitary wave, which is a single crest of water above the still-water level traveling without change of form at a constant speed, with a net displacement of water in direction of wave travel. It is further characterized by its independence from wave length.

Fig. 28-1 shows the oscillatory wave form and its characteristics. In deep water, each particle of water on the wave surface describes a circle, the radius of which is one-half the wave height about its normal center, midway between the crest and trough of the wave. The center line of rotation is elevated above the still-water level by height h_0 because the crest is at a greater distance above still water than is the trough below it. The difference depends upon the wave steepness; for a very steep wave the proportion is about two-thirds above and one-third below the still-water level. At any instant the small arrows on the circular paths indicated by dotted circles in Fig. 28-1 show the relative position of the particles and their direction of motion in the formation of the wave. The heavy line connecting the arrows is the wave form at the surface of the water at that particular instant, the length between two consecutive crests is the wave length L, and the height between the trough and the crest is the wave height or *amplitude*(振幅) H. The wave form travels over the water surface, and the time for two consecutive crests to pass a point is the wave period T. The speed of the wave form is called the wave velocity or velocity of wave propagation.

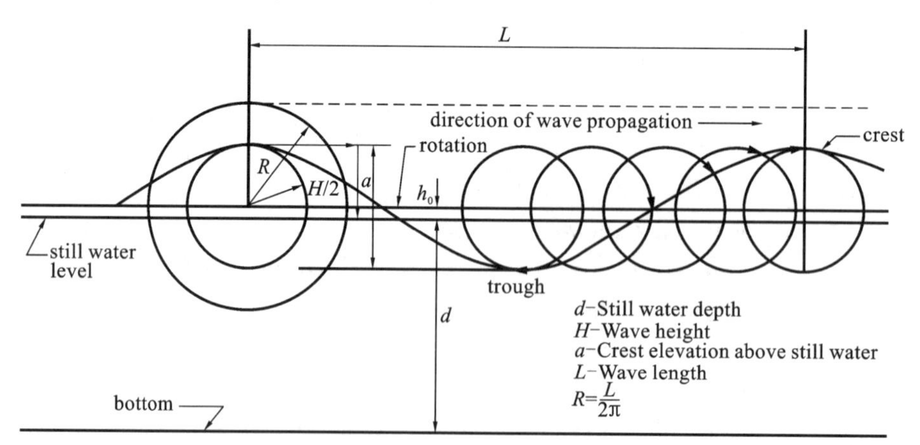

Fig. 28-1 Deep-water wave characteristics and form

科技英语写作技巧(15)——论文摘要(Ⅲ)

下面是一个土木工程专业论文摘要的实例。

This paper presents the *results* of an analysis of the global structural behaviour of the 8-storey steel framed building at Cardington *during* the two BRE large-scale fire *tests*. These tests were *carried out to investigate* the performance of the whole building structures under realistic fire conditions and to *provide* quality *experimental information for* the *validation* of various numerical models. The *results* of these two BRE tests were *analysed* using a specialist finite element computer program (to be referred to as FIREFRAME in this paper) and this paper *presents the main findings* of this study. The *results* of this analysis *seem to indicate* that FIREFRAME is capable of simulating flexural bending behaviour. However, in order for the program to simulate the Cardington frame behaviour during the Corner test, it *requires* a more advanced numerical procedure *to deal with* slab tensile membrane behaviour at large deflections. Test and computer simulation *results suggest* that columns may attain large moments as a result of being pushed by the adjacent hot beams, but as the test column temperatures were low, it was not possible to assess the column failure behaviour. *Furthermore*, computer simulations *indicate* that large sagging moments may develop in heated beams during cooling, but *further research is required* to check whether this would lead to beams failure.

该摘要对论文的要点进行概括，提出了试验方法并对今后的研究提出了建议，其所涉及的是论文的主题、研究中所采用的方法、主要研究结果及数据、研究的结论和对开展下一步研究工作的建议等内容。读者通过阅读摘要即可大概了解论文的主要内容。通常，这种摘要反映了论文的基本面貌，能够代替阅读论文全文。读者在阅读上例时请注意斜体部分的表达，可以学习借鉴。

Lesson 29

HARBOR PLANNING

A harbor is a water area partially and so protected from storm as to provide safe and suitable accommodation for vessels seeking refuge, supplies, refueling, repairs, or the transfer of cargo.

Harbors may be classified as natural, seminatural, or artificial, and as harbors of refuge, military harbors, or commercial harbors. Commercial harbors may be either municipal or privately owned.

A port is a sheltered harbor where marine terminal facilities are provided, consisting of piers or wharves at which ships berth while loading or unloading cargo, transit sheds and other storage areas where ships may discharge incoming cargo, and warehoused where goods may be stored for longer periods while awaiting distribution or sailing. ① Thus the terminal must be served by railroad, highway, or inland-waterway connections, and in this respect the area of influence of the port reaches out for a considerable distance beyond the harbor. The tributary area of a port consists of that portion of the adjacent area for which freight transportation costs are lower than they are to competing ports. The opening of the Saint Lawrence River Seaway has resulted in completely new tributary areas for many of the Great Lakes port for certain commodities. The harbor then is a very important comprehensive system of works and services which comprise a port.

The decision to build a port, and its location, generally will be determined by factors having to do with its need and economic justification, prospective volume of seaborne commerce, and availability of inland communications by both land and water.

Assuming that above studies have been made and the general location of the harbor has been established, as well as its principal use and the type and tonnage of traffic to be handled, the next step, which in some cases will have been initiated during the above studies, will be to make preliminary studies and layouts of the port in preparation for making a complete site investigation to gather all the information which will be needed in making the final design of the port.

With the general requirements of the port having been established and preliminary site information obtained, the next step will be to make preliminary studies of harbors and port layouts, which will usually be supplemented with approximate cost estimates based on certain assumptions which will have to be verified when the site investigation is made. [2] This preliminary planning will include the following.

Determining best location of a harbor. Unless the site is fixed by specific requirements of the port, several locations of the harbor will have to be studied, to determine the most protected location involving the least amount of dredging and with the most favorable bottom conditions as well as a shore area suitable for the development of the terminal facilities.

Size and shape of a harbor and turning basin. The number and size of ships using a harbor will determine its size to a large extent, but existing site conditions will also have an important influence. Generally speaking, unless the harbor is a natural one, its size will be kept as will permit safe and reasonably comfortable operations to take place.

Type, location, and height of breakwaters. Breakwaters are required for the protection of artificial and semi-natural harbors. Their location and extent will depend upon the minimum size of harbor required for the anticipated traffic in the port.

Location and width of entrance to harbor. In order to reduce the wave height within the harbor, entrances should be no wider than necessary to provide safe navigation and to prevent dangerous currents when the tide is coming in and going out. The entrance width should be in proportion to the size of the harbor and the ships using it.

Depth of harbor and approach channel. For ideal operating conditions the water in the approach channel, in the entrance, and in the harbor should be of sufficient depth to permit navigation at lowest low water when the ship is fully loaded.

Number, location, and type of docks. Docking facilities vary widely from port to port. They may consist of a single pier or as many as a thousand piers in the Port of New York. The number of berths will depend upon the anticipated number of ships to use the port and the time it will take to discharge and take on cargo or passengers.

Words and Expressions

harbor	[ˈhɑːbə]	*n.* 海港
vessel	[ˈvesəl]	*n.* 船，容器
refuge	[ˈrefjuːdʒ]	*n.* 庇护，避难，避难所
refuel	[ˌriːˈfjuəl]	*v.* 补给燃料
cargo	[ˈkɑːgəu]	*n.* 船货，(车、船、飞机等运输的)货物
seminatural	[ˌsemiˈnætʃərəl]	*adj.* 半自然的
port	[pɔːt]	*n.* 港口，舱门，左舷，避风港

pier	[piə]	*n.* 凸式码头,凸岸码头
wharf(*pl.* wharves)	[(h)wɔ:f]	*n.* 码头,顺岸码头
		v. 卸上码头
berth	[bə:θ]	*n.* 停泊处
		v. 使停泊
warehouse	[ˈwɛəhaus]	*n.* 仓库,货栈
seaway	[ˈsi:wei]	*n.* 海上航道
justification	[dʒʌstifiˈkeiʃ(ə)n]	*n.* 正当的理由,辩护
seaborne	[ˈsi:bɔ:n]	*adj.* 海上运输的,漂流的
tonnage	[ˈtʌnidʒ]	*n.* 吨位,排水量
preliminary	[priˈliminəri]	*adj.* 预备的,初步的
breakwater	[ˈbreikˌwɔ:tə (r)]	*n.* 防浪堤
dock	[dɔk]	*n.* 码头,船坞
		v. 使(船)靠码头
harbor of refuge		避风港
inland-waterway		内陆水道,内河
tributary area of a port		港口腹地
turning basin		回转水域
artificial harbors		人工港
semi-natural harbors		半天然港

Notes

① A port is a sheltered harbors where marine terminal facilities are provided, consisting of piers or wharves at which ships berth while loading or unloading cargo, transit sheds and other storage areas where ships may discharge incoming cargo, and warehoused where goods may be stored for longer periods while awaiting distribution or sailing.
港口是一个可以提供掩护作用的海港,它提供作为海运基地的作业设施,并包括若干凸岸或顺岸码头,可供船舶停靠装卸货物,有前方仓库或其他货场供船只卸存货物,也有后方仓库或堆场供货物较长时期存放,等待分配及运出。

② With the general requirements of the port having been established and preliminary site information obtained, the next step will be to make preliminary studies of harbors and port layouts, which will usually be supplemented with approximate cost estimates based on certain assumptions which will have to be verified when the site investigation is made.
对于港口的一般要求既经确定并已获得有关港址的初步原始资料以后,下一步工作应对港口水域和港埠平面布置进行初步研究。所做的研究通常应提供基于一定假设的大致成本估算,而这些假设必须要在将来进行的港址调查中被验证。

Exercises

Ⅰ. Translate the following words into Chinese.

1. tributary area of a port ______________________________
2. semi-natural harbors ______________________________
3. harbor of refuge ______________________________
4. seaway ______________________________
5. mouths of rivers ______________________________

Ⅱ. Translate the following words into English.

1. 内河港口 ______________________________
2. 海运作业区 ______________________________
3. 航道边界线 ______________________________
4. 避风港 ______________________________
5. 护岸 ______________________________

Ⅲ. Translate the following sentences into Chinese.

1. A harbor is a water area partially and so protected from storm as to provide safe and suitable accommodation for vessels seeking refuge, supplies, refueling, repairs, or the transfer of cargo.
2. The tributary area of a port consists of that portion of the adjacent area for which freight transportation costs are lower than they are to competing ports.
3. The number and size of ships using a harbor will determine its size to a large extent, but existing site conditions will also have an important influence.

Ⅳ. Translate the following sentences into English.

1. 回转水域有不同的形状，要依据港口规模、船舶泊位的数量及其排列而定。
2. 如经济条件允许，港口还应为沿海航行的小型船只留备泊地，供它们在码头等待泊位时和在恶劣气候时作掩护之用。
3. 为了港口发展，港口陆域所需的面积应满足一切货物的装卸和储存，以及旅客交通的需要。
4. 港口作为船货装卸的作业场地，必须有宽度适当的码头前沿地带，后面则为码头前方仓库或临时货棚。
5. 在港口的客运区，应于水边或临近水边处留出较大的场地，以适应旅客及游览者的车辆需要。

Ⅴ. Answer the following questions briefly.

1. How many types of harbors are there?
2. What is the purpose of ports?
3. How is the location of harbor determined?

Reading Materials

Ports

Ocean ports are usually located in natural harbors in bays, tidal estuaries, and river mouths, or they may be formed on an unprotected shore line by the construction of breakwaters. In a broader sense and from the standpoint of commerce and foreign trade, they are any ports of call which can be reached directly by large ocean-going vessels. In this respect they may be located several hundreds of miles up rivers or on lakes, although from a harbor engineering viewpoint such are usually considered to be river or lake ports. The opening of the Saint Lawrence Seaway, for instance, has in effect created ocean ports out of such lake ports as Buffalo, Cleveland, and Chicago. Philadelphia and New Orleans are many miles from the ocean on large rivers with maintained deep-water channels, yet they are thought of in terms of ocean rather than rive ports. The port of Houston, Texas, located some 50 miles inland, is a striking example of the artificial creation of an ocean port by means of dredging a deep-draft ship canal to the ocean.

A free port or zone is an isolated, enclosed, and policed area in or adjacent to a port of entry, without a resident population. Furnished with the necessary facilities for loading and unloading, for supplying fuel and ship's stores, for storing goods and reshipping them by land and water, it is an area within which goods may be landed, stored, mixed, blended, repacked, manufactured, and reshipped without payment of duties and without the intervention of customs officials. The purpose of the free zone is to encourage and expedite that part of a nation's foreign trade which its government wishes to free from the restrictions necessitated by customs duties. The most important free port in Europe is Hamburg, which was originated about 1883 and has grown ever since.

A marine terminal is that part of a port or harbor which provides *docks*(船坞), cargo-handling, and storage facilities. When only passengers embark and disembark along with their baggage and miscellaneous small cargo generally from ships devoted mainly to the carrying of passengers, it is called a passenger terminal. When the traffic is mainly cargo carried by freighters, although many of these ships may carry also a few passengers, the terminal is commonly referred to as a freight or cargo terminal. In many cases it will be known as a bulk cargo terminal, where such products as petroleum, cement, and grain are stored and handled. For instance, a refinery will usually have a storage-tank farm, pipelines for transporting the crude oil and refined products, and docking facilities for *mooring*(停泊) the tankers while they are being loaded and unloaded; all these comprise the marine terminal of the refinery.

An offshore mooring is provided usually where it is not feasible or economical to construct a dock or provide a protected harbor. Such an anchorage will consist of a number of anchorage units, each consisting of one or more anchors, chains, sinkers, and buoys to which the ship will attach its mooring lines. These will be supplemented in most cases by the ship's bow anchors. Bulk cargo is usually transported to or from the ship by pipeline or trestle conveyor, while other cargo may be transferred by lighter.

An anchorage area is a place where ships may be held for *quarantine*(检疫，隔离) inspection, await docking space while sometimes removing ballast in preparation for taking on cargo, or await favorable weather conditions. Special anchorages are sometimes provided for ships carrying explosives or dangerous cargo and are usually so designated on harbor maps by name and depth of water. Anchorages are usually located away from the marine terminal and adjacent to main channels so as to be near deep water but out of the path of the main traffic. They may be in naturally protected areas or protected from wind and waves by breakwaters.

A turning basin is a water area inside a harbor or an enlargement of a channel to permit the turning of a ship. When space is available, the area should be at least twice the length of the ship to permit either free turning or turning with the aid of tugs, if wind and water conditions require. When space is limited, the ship may be turned by either warping around the end of a pier or turning dolphin, either with or without the use of its lines, and the turning basin will be much smaller and of a more triangular or rectangular shape.

科技英语写作技巧(16)——论文标题

论文标题的基本功能是概括全文，除此之外，还应具备吸引读者和便于检索等特点。

1. 论文标题的语言特点。

(1)一般不用完整的句子，不包含主语、谓语、宾语、状语等句子成分。

(2)多用名词、名词词组或者动名词。

2. 写作要求。

(1)简短。

论文的标题力求言简意赅，一般控制在10个词以内，最多不超过15个词。有人根据人们记忆的特点，提出最好控制在12个词以内。如果标题太长，可作为副标题处理。当然，过于简短而令人费解的题目也不可取。

(2)不加冗余套语。

英文论文标题中一般可以不用加：regarding…, studies (investigation/ observation) on…, the method of…, some thoughts on…, a research on…等套语。由于文

章性质不言而喻,这些词语大多数可以省略。

(3)少用问题性标题。

例如:Is there…? When does…? Should the…?

常用格式为:中心词+后置定语,例如:比较(comparison of)、影响(effects of)、分析(analysis of)、研究(research on)等。

(4)避免名词与动名词混杂使用。

论文标题中的并列内容,应该在语法结构上对称,或者名词和名词并列,或者动名词和动名词并列。

Lesson 30

WHARVES AND PIERS

A dock is a general term used to describe a marine structure for the mooring or tying up of vessels, for loading and unloading cargo, or for embarking and disembarking passengers. More specifically a dock is referred to as a pier, wharf, bulkhead, or, in European terminology, a jetty, quay, or quay wall. In Europe, where there are large variations in tide level, it is commonly known as an artificial basin for vessels and is called a wet dock. When the basin is pumped out, it is termed a dry dock.

A wharf or quay is a dock which parallels the shore. It is generally contiguous with the shore, but may not necessarily be so. On the other hand, a bulkhead or quay wall, while similar to a wharf and often referred to as such, is backed up by ground, as it derives its name from the very nature of holding or supporting ground in back of it. ①

A pier or jetty is a dock which projects into the water. Sometimes it is referred to as a mole and in combination with a breakwater which is termed a breakwater pier. As contrasted to a wharf, which can be used for docking on one side only, a pier may be used on both sides, although there are instances where only one side is used owing to either the physical conditions of the site or the lack of need for additional berthing space. ②

The construction of wharves, piers, bulkheads, and fixed mooring berths falls generally into two broad classifications: ① docks of open construction with their decks supported by piles or cylinders, and ②docks of closed or solid construction, such as sheet-pile cells, bulkheads, cribs, caissons, and gravity (quay) walls.

Docks of open construction may be further subdivided into what are commonly referred to as ①high-level decks and ②relieving-type platforms in which the main structural slab is below the finished deck and the space between is filled to provide additional weight for stability, as shown in Fig. 30-1.

Sheet-pile bulkheads may be constructed of wood, steel, or concrete sheet piling which may be supported by tie rods attached to an anchor wall or anchor piles located a safe distance in back of the face of the bulkhead, as shown in Fig. 30-2, or by batter piles along the rear of the piling, as shown in Fig. 30-3. In shallow installations where the bottom is of good supporting value, the sheet piling may be driven deep enough to act as a cantilever without the benefit of additional support. ③

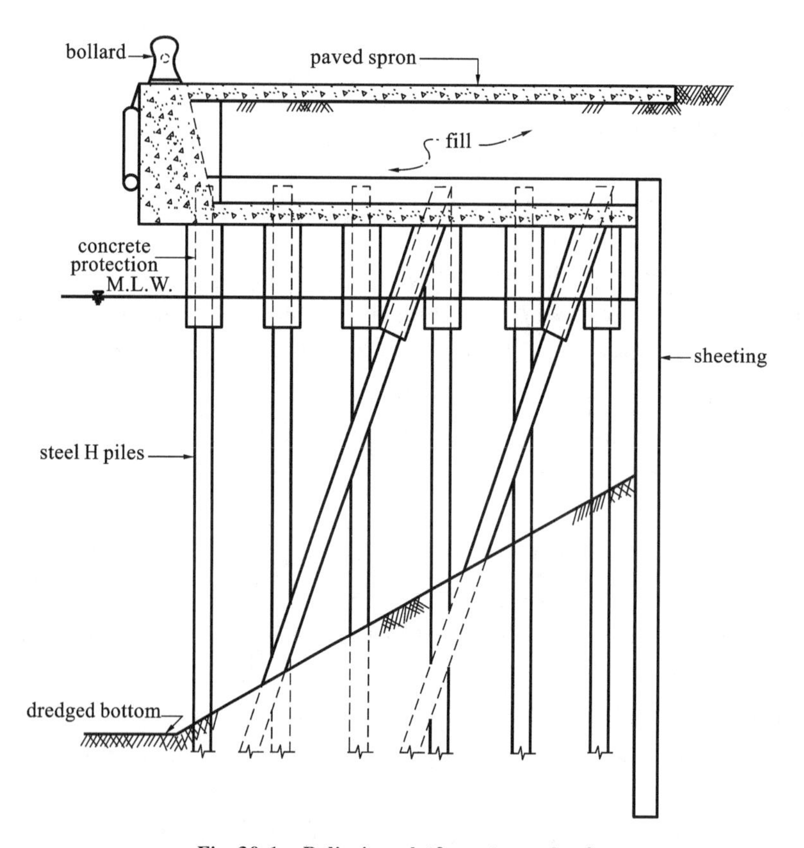

Fig. 30-1 Relieving-platform-type wharf

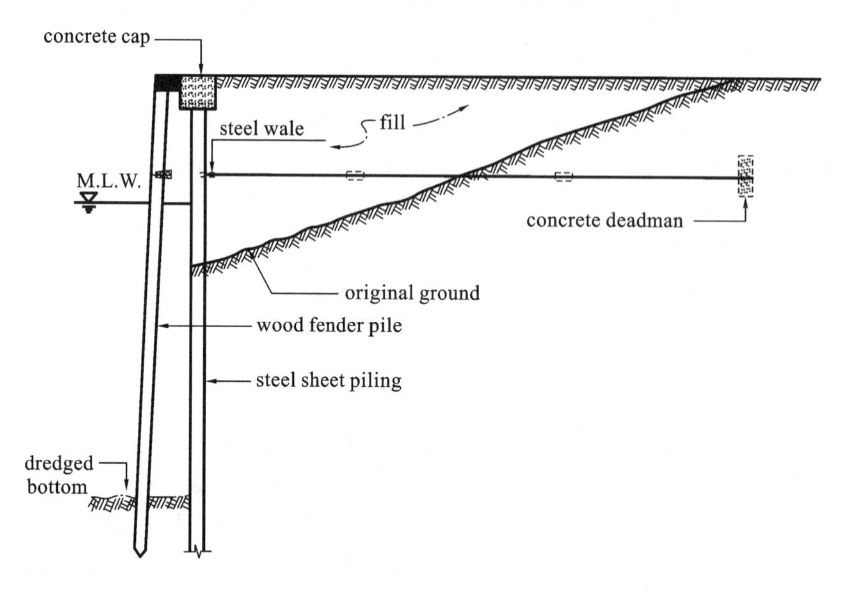

Fig. 30-2 Sheet-pile bulkhead supported by tie rods and anchor wall

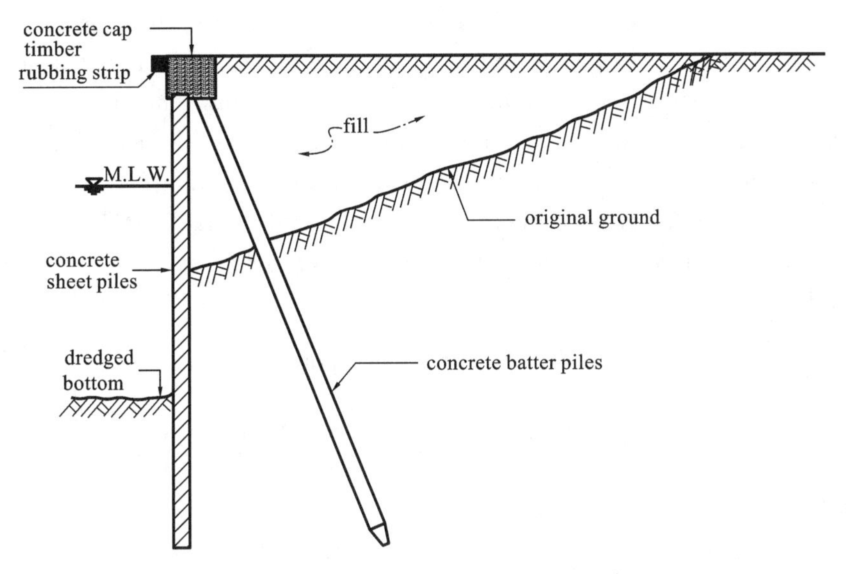

Fig. 30-3 The sheet-pile bulkhead supported by batter piles

Concrete caissons have been used quite extensively for the construction of wharves or quay walls, especially in Europe. A striking example of this type of construction is the 7500-ft-long quay built at Southampton, England, in 1934, which consists of 146 concrete caissons sunk to from 7 ft to 30 ft below dredged bottom. Caissons may have open wells and cutting edges so that they may be sunk below the dredged bottom in order to obtain a firm support, or they may have a closed bottom, in which case they are lowered on to a prepared bottom usually consisting of a gravel or crushed-stone bed or leveling course, as shown in Fig. 30-4. The latter type of caisson is usually filled with rock or granular material so as to

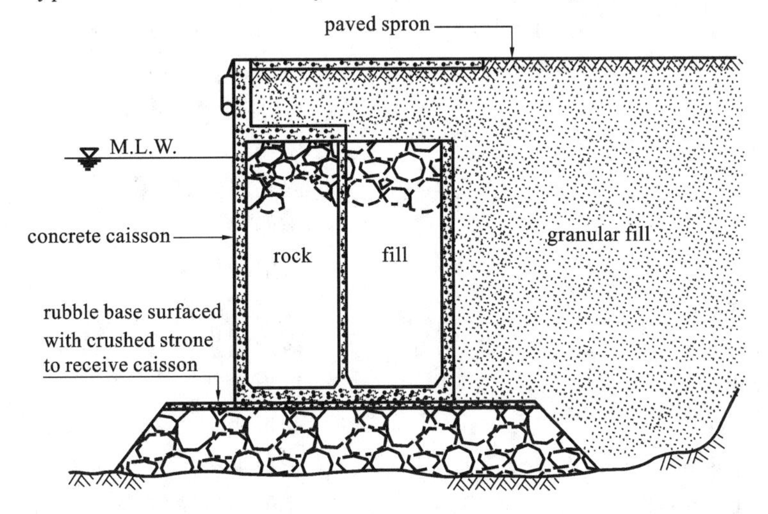

Fig. 30-4 The solid-type bulkhead wharf using concrete caissons with closed bottom

provide added weight for stability. Caissons are generally constructed of such height that their tops will be at or a little above low-water level and are surmounted by a gravity dock wall of cast-in-place concrete. This permits the upper part of the dock face to be constructed to true alignment and grade, as well as enabling provision to be made for the attachment of the fender system, backing log, etc., and the installation of railroad track, crane rails, utilities, and foundations, if required, in the fill above the caissons and in back of the dock wall or parapet, before laying the apron paving.

Words and Expressions

marine	[məˈriːn]	*n.* 船舶,海景画
mooring	[ˈmuəriŋ]	*n.* 停泊处
embark	[imˈbɑːk]	*v.* 上船,从事,发动
bulkhead	[ˈbʌlkhed]	*n.* 隔板,防水壁,驳岸
terminology	[ˌtəːmiˈnɔlədʒi]	*n.* 术语学
jetty	[ˈdʒeti]	*n.* 码头,防波堤
quay	[kiː]	*n.* 码头
mole	[məul]	*n.* 防波堤,筑有防波堤的海港
cylinder	[ˈsilində]	*n.* 圆筒,圆柱体,柱面
crib	[krib]	*n.* 栅栏
caisson	[ˈkeisən]	*n.* 沉箱(桥梁工程)
slab	[slæb]	*n.* 厚平板,混凝土路面,板层
rear	[riə]	*n.* 后面,背后,后方,屁股
granular	[ˈgrænjulə]	*adj.* 颗粒的,粒状的
surmount	[səːˈmaunt]	*v.* 战胜,超越,克服,在……顶上
parapet	[ˈpærəpit]	*n.* 女儿墙,胸墙
apron	[ˈeiprən]	*n.* 遮檐板,窗台
quay wall		岸墙
wet dock		湿船坞,系船渠
dry dock		干船坞
breakwater pier		防浪堤码头
solid construction		实体式结构
sheet-pile cell		格形板桩
gravity wall		重力式墙
high-level deck		梁板式
relieving-type platform		减载平台
sheet-pile bulkhead		板桩驳岸

tie rod	拉杆
batter pile	斜桩
concrete caisson	混凝土沉箱
dredged bottom	挖成的水底
leveling course	找平层
crane rail	起重机铁轨

Notes

① On the other hand, a bulkhead or quay wall, while similar to a wharf and often referred to as such, is backed up by ground, as it derives its name from the very nature of holding or supporting ground in back of it.
反之,与顺岸码头类似而称为驳岸或岸壁墙的建筑物,则以地面为支撑,两者都因它后面有地面这一共性而得名。

② As contrasted to a wharf, which can be used for docking on one side only, a pier may be used on both sides, although there are instances where only one side is used owing to either the physical conditions of the site or the lack of need for additional berthing space.
凸岸码头与只能单侧靠船的顺岸码头不同,它可以两侧靠船,虽然有些凸岸码头由于场地条件所限,或因不需要更多泊位,而只单侧靠船。

③ In shallow installations where the bottom is of good supporting value, the sheet piling may be driven deep enough to act as a cantilever without the benefit of additional support.
在水深小而地基承载力强的驳岸,可将板桩打入足够深的地基,使其受力情况与悬臂梁相似而不另加支撑。

Exercises

Ⅰ. Translate the following words into Chinese.

1. concrete caissons ________________
2. rock-mound breakwaters ________________
3. wet dock ________________
4. armor course ________________
5. concrete or steel sheet-pile walls ________________

Ⅱ. Translate the following words into English.

1. 进港航道 ________________
2. 锚地 ________________
3. 透空式结构 ________________
4. 斜桩锚固 ________________
5. 减载平台 ________________

Ⅲ. **Translate the following sentences into Chinese.**

1. In Europe, where there are large variations in tide level, it is commonly known as an artificial basin for vessels and is called a wet dock.
2. Concrete caissons have been used quite extensively for the construction of wharves or quay walls, especially in Europe.
3. The latter type of caisson is usually filled with rock or granular material so as to provide added weight for stability.

Ⅳ. **Translate the following sentences into English.**

1. 在岸线附近,海底地质条件可能比较好,那么建造顺岸码头是有利的。
2. 码头设计应考虑的问题包括:第一,码头或顺岸码头的尺度和布置;第二,设计采用的轮廓形式;第三,设计采用的荷载。
3. 堆石防波堤可以经受相当大的沉降量,因为它的结构性质就是允许发生内部调节而不影响其总体的强度。
4. 凸岸码头可以或多或少地与岸线平行,用凸堤与岸上联系。
5. 块石是建造防波堤的主要材料之一,故对于石料的可用性应予以调查研究。

Ⅴ. **Answer the following questions briefly.**

1. What is the concrete caisson? Try to state it in your own words.
2. What is the purpose of docks?
3. What may docks of open construction be further subdivided into?

Reading Materials

Breakwaters

A *breakwater*(防波堤) is a structure constructed for the purpose of forming an artificial harbor with a water area so protected from the effect of sea waves as to provide safe accommodation for shipping. There are two classes of breakwater: those giving protection to commercial harbors or their entrances and those sheltering an anchorage or roadstead, being used by vessels to escape the violence of storms or while awaiting orders and their turn to dock. Such an anchorage may be an outer harbor where there are no docks.

There are many different types of breakwaters which have been constructed in all parts of the world. Natural rock and concrete or a combination of both are the materials which form 95 percent or more of all breakwaters constructed. Steel, timber, and even compressed air have served to a lesser extent to break the force of the sea waves.

Most breakwaters function only to provide protection, but occasionally they serve a dual purpose by becoming part of a pier or supporting a roadway. The former is termed a breakwater pier or *quay*(码头) and the latter a mole.

There are two main types of breakwaters, the *mound*(土堆) type and the wall type. The breakwaters are constructed by the following: ① natural rocks, ② concrete blocks, ③ a combination of rocks and concrete blocks, and ④ concrete tetrapods and tribars. These types of breakwaters may be supplemented in each case by concrete monoliths or seawalls to break the force of the waves and to prevent splash and spray from passing over the top. In the second main classification of breakwater there are such types as: ① concrete-block gravity walls, ② concrete caissons, ③ rock-filled sheet-pile cell, ④ rock-filled timber cribs, and ⑤ concrete or steel-pile wall.

The type of breakwater to be used is usually determined by the availability of materials at or near the site, the depth of water, the condition of the sea bottom, its function in the harbor, and, last but not the least, the equipment suitable and available for its construction.

Since the main purpose of the breakwater is to provide protection from the waves, it follows that an understanding of wave action and its force is one of the more important elements in its design. Another important element is the character of the waves against the structure which has been placed to *dissipate*(消耗,耗散) the energy of the waves. In this respect, since most breakwaters are gravity structures, they depend upon their weight for stability. Therefore, the depth of water and the character of the bottom are important factors in their design. Practical considerations usually limit the height of vertical-type breakwaters to a water depth of not over 60 ft below mean sea level, and in deeper water they generally found on a rock fill below this level; otherwise the width of the structure becomes unwieldy. The character of the bottom may well be the determining factor in the type of the breakwater selected, as it is usually difficult, if not impossible, and expensive to prepare a solid foundation on soft material for the support of a wall-type gravity structure.

Since rock is one of the main materials used in the construction of breakwaters, its availability will have to be investigated. In this respect not only will it be necessary to determine that it is economically feasible to produce and deliver to the site a sufficient quantity of rock, but its density, soundness, and ability to break into large pieces when *quarried*(采石) will be important factors in determining its use.

The abundance of durable rock and familiarity with quarrying methods to produce large quantities of rock at economical costs have led to the adoption of rock-mound breakwaters to a greater extent than any other type for the protection of harbors along the North America and South America seacoasts. The first breakwater ever constructed in the United States, as shown in Fig. 30-5, is the Delaware Breakwater near Cape Henlopen, Delaware, a rock mound which was commenced in 1828, but not completed until 1869. It is 5267 ft long and has served as a prototype for other mound breakwaters.

On the Great Lakes the early practice was to use large amounts of timber because of its availability and cheapness, but in recent years rock and concrete have been used more extensively.

Continental practice, particularly in Italy, has greatly favored the vertical-wall type, supplemented in very deep water by a *rubble*(碎石) base. Extreme depths of water, often over 100 ft, and great wave heights have in many cases brought about this general practice. Experience over many years in using this type of construction for docks and other marine structures, resulting in confidence and familiarity in the casting and placing of large concrete blocks and monoliths, either hollow or solid, and the more common use of a combination breakwater and dock, or quay, have contributed to the use of this type of design in European practice.

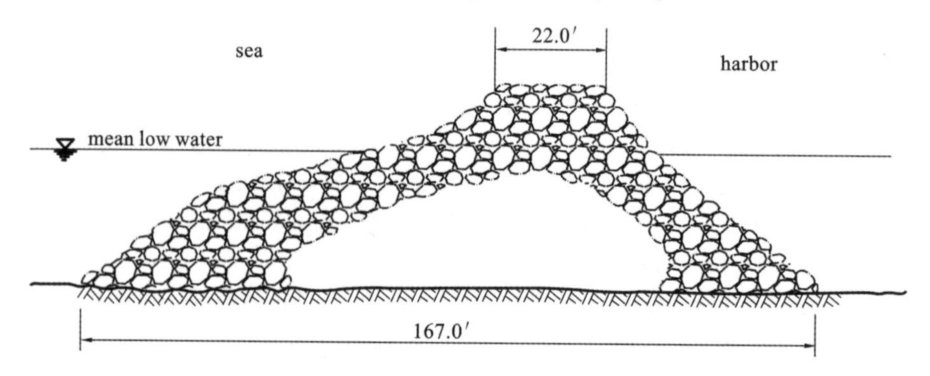

Fig. 30-5 First rock-mound breakwater constructed in the United States: the Delaware Breakwater

科技英语写作技巧(17)——关键词与作者姓名

1. 关键词(key words)。

按照惯例,摘要的末尾要有关键词,表明文章的特性,数量一般为 3～10 个。

关键词也称主题词,它是为了检索的需要,从论文中选出的最能代表论文中心内容特征的词或词组。有了关键词就便于检索刊物编制索引和输入计算机检索系统,方便刊登该文的刊物编制年终索引,同时也有助于读者了解该文的主题及编排个人检索卡片。

关键词是论文信息最高度的概括,是论文主旨的概括体现,因此必须准确、恰当。

关键词的选择大多从标题和摘要中产生,每个关键词与下一个关键词用分号分开,最后一个没有标点符号,每个词或词组中的任何一个实词首字母一般要大写。

2. 作者姓名(author's name)。

作者的姓名一定要用全称(full name)。这主要是针对外国人来说的,因为他们的姓名一般是由三部分(即:教名——Christian name, first name, given name 或

forename；中间名——middle name；姓——surname）构成的，有的只有两部分，有的可由四部分构成。中国人的姓名应根据我国规定按汉语顺序用汉语拼音文字写出，如："李伟东"应写成 Li Weidong（也有不少人仍采用韦氏拼写法）。若为了防止外国人误解，可以在姓的下面画一条横线，或姓全用大写字母，或在姓后用逗号与名分开，使国外杂志的编辑人员知道这是你的姓。也有不少人仍按外国人的姓名顺序写法表示成 Weidong Li，Weidong-Li，甚至写成 WeiD. Li 或 W. D. Li。

附　　录

土木类常用英文期刊数据库

El Village
Web of Science(SCI)
ASCE
Elsevier Science
John Wiley & Sons Springer-LINK
PQDD(B)

EI 收录的常用国内土木类期刊(部分)

土木工程学报
建筑结构学报
岩石力学与工程学报
中国公路学报
岩土力学
力学学报
工程力学
计算力学学报
振动工程学报
应用力学学报
振动与冲击
基础科学与工程学报
清华大学学报（自然科学版）
同济大学学报（自然科学版）
东南大学学报（自然科学版）
哈尔滨工业大学学报
天津大学学报
湖南大学学报（自然科学版）
华南理工大学学报(自然科学版)
华中科技大学学报(自然科学版)
上海交通大学学报
中南大学学报(自然科学版)

参考文献

[1] TONIAS D E, ZHAO J J. Bridge engineering-design, rehabilitation, and maintenance of modern highway bridges [M]. 2nd edition. New York: McGraw-Hill, 2002.

[2] BARKER R M, PUCKETT J A. Design of highway bridges: an LRFD approach[M]. 2nd edition. Hoboken: John Wiley & Sons, 2006.

[3] HARRIS R, KELLY O. The structural engineering of the downland gridshell, space structures[M]. London: Thomas Telford , 2002.

[4] SMITH B S, COULL A. Tall building structures: analysis and design[M]. New York: John Willey& Sons, Inc, 1998.

[5] DYRBYE C, HANSEN S O. Wind loads on structures[M]. Chichester: John Wiley& Sons, 1999.

[6] WIGHT J K, MACGREGOR J G. Reinforced concrete—mechanics and design[M]. New Jersey: Prentice Hall, 1988.

[7] SCHOLOCK D L. Structures[M]. 5th Edition. New Jersy: Pearson Prenston Hall, 2004.

[8] MACCINLEY T J, ANG T C. Structural steelwork:design to limit state theory[M]. Norwich: Butterworth, 1987.

[9] Geotechnical Engineering Office, Civil Engineering and Development Department, the Government of the Hong Kong Special Administrative Region. Foundation design and construction[M]. GEO Publication. No. 1/2006.

[10] BOWLES J E. Foundation analysis and design[M]. 5th Edition Columbus: McGraw-Hill, 1997.

[11] WICANDER R, MONROE J S, MCLODAW S. Essentials of geology[M]. Florence: Wadsworth, 1998.

[12] MASAHIKO K, MIKIO S. The principles of construction management[M]. Tokyo: Sankaido, 1996.

[13] CHANG L M. The trends of construction IT in the USA[C]. AN X H. Innovative application of information technology in construction. Beijing: Tsinghua University Press, 2004.

[14] QUINN A D. Design and construction of ports and marine structures[M]. New York: McGraw-Hill, 1972.

[15] GHALI A, NEVILLE A M, BROWNT G. Structural analysis:a unified classical and matrix approach[M]. 6th Edition. London: Chapman and Hall, 2009.

[16] MCCORMAC J C,BROWN R H. Design of reinforced concrete[M]. New

York: John Willey& Sons, 2001.

[17] MCCORMAC J C, CSERNAK S F. Structural steel design[M]. 5th Edition. New York: Happer - Collins, 1992.

[18] NILSON A H, DARWIN D, DOLAN C W, et al. Design of concrete structures[M]. 张川,缩编. 重庆:重庆大学出版社, 2005.

[19] 刘澜. 交通运输专业英语[M]. 成都:西南交通大学出版社, 2006.

[20] 李亚东. 新编土木工程专业英语[M]. 成都:西南交通大学出版社, 2000.

[21] 魏汝尧,董益坤. 科技英语教程[M]. 北京:北京大学出版社, 2005.

[22] 陈权,王丽娟. 科技英语[M]. 北京:中国铁道出版社, 2005.

[23] 惠宽堂,曹勇. 土木工程专业[M]. 北京:中国建筑工业出版社, 2006.

[24] 王建武,李民权,曾小珊. 科技英语写作——写作技巧、范文[M]. 西安:西北工业大学出版社, 2000.

[25] 丁西亚. 英语科技论文写作——理论与实践[M]. 西安:西安交通大学出版社, 2006.

[26] 秦荻辉. 科技英语写作教程[M]. 西安:西安电子科技大学出版社, 2001.

[27] 胡庚申. 怎样进行国际交流——国际交流"How-to"电视讲座[M]. 北京:北京工业大学出版社, 1992.

[28] 钱永梅,庞平. 新编土木工程专业英语(建筑工程方向)[M]. 北京:化学工业出版社,2004.

[29] 沈祖炎. Introduction of civil engineering[M]. 北京:中国建筑工业出版社, 2005.

[30] 段兵廷. 土木工程专业英语[M]. 2 版. 武汉:武汉理工大学出版社, 2003.

[31] 王霞. 英语写作中母语负向迁移的实例分析及对策[J]. 教书育人, 2012(36): 83-85.

[32] STEVE W. 如何成为学术论文写作高手[M]. 北京:北京大学出版社,2015.